Understanding Mathematics for Aircraft Navigation

Understanding Mathematics for Aircraft Navigation

James S. Wolper

McGraw-Hill

New York Chicago San Francisco Lisbon London Madrid
Mexico City Milan New Delhi San Juan Seoul
Singapore Sydney Toronto

Library of Congress Cataloging-in-Publication Data

Wolper, James S.
    Understanding mathematics for aircraft navigation / James S. Wolper.
       p. cm.
    Includes bibliographical references and index.
    ISBN 0-07-137572-4
    1. Navigation (Aeronautics)—Mathematics. I. Title.

    TL586.W65 2001
    629.132′51′0151—dc21   2001030508

McGraw-Hill

A Division of The McGraw·Hill Companies

2 3 4 5 6 7 8 9 0 DOC/DOC 0 7 6 5

ISBN 0-07-137572-4

The sponsoring editor for this book was Shelley Ingram Carr, the editing supervisor was Sybil M. Sosin, and the production supervisor was Sherri Souffrance. It was set in Slimbach by Thomas Technology Solutions, Inc.

Printed and bound by R. R. Donnelley & Sons Company.

McGraw-Hill books are available at special quantity discounts to use as premiums and sales promotions, or for use in corporate training programs. For more information, please write to the Director of Special Sales, Professional Publishing, McGraw-Hill, Two Penn Plaza, New York, NY 10121-2298. Or contact your local bookstore.

 This book is printed on recycled, acid-free paper containing a minimum of 50% recycled, de-inked fiber.

I began work on this book around 1984, and since that time I have learned from many pilots, students, and professors: I sincerely thank them all, even if not by name. Mark Peterson of Mount Holyoke College first exposed me to the idea of using vector methods in spherical geometry (in the context of Dante's *Inferno*). Jim Mansfield shared his experiences crossing oceans in airplanes ranging from the DC-4 to the Boeing 747. Harriet Pollatsek of Mount Holyoke and Larry Ford of Idaho State University encouraged my academic use of this material, and Shamim Mohamed, Jerry Priddy, and Scott Hughes were sharp flying students with technical backgrounds who, perhaps unknowingly, forced me to take care in my technical explanations. Allan Rifkin and René Schmauder helped me with some early flight experiments, and I enjoyed many hours of discussion with them.

My wife Terri Ross and my children Ellie and Nate have given me time to work and tolerated my navigational experiments on our family vacations. I fondly recall the night when Terri understood what I was after. We were crossing South Dakota in a small plane, and I set myself the task of flying across the state "direct" while on an IFR flight plan, without using my GPS. The kids slept peacefully in back while I worked with charts, knobs, and computers, keeping us on course. She turned to me, smiling, and asked: "You really enjoy this, don't you?" Thanks.

Similarly, I thank my fellow pilots at the Avcenter. I can't remember anyone getting upset when I said, "Hey, how about we try . . ." but this was usually on a freight flight in the middle of nowhere in the middle of the night, so maybe I couldn't see them rolling their eyes. I hope I haven't been too much like Captain Dudley in (Gann). Our principal operations inspector, Jim Ralph of the FAA, has always encouraged me in my search to learn more about the art and science of navigating.

I also acknowledge the support of Mount Holyoke College, Idaho State University, the U.S. Air Force Office of Scientific Research, and the NASA–Idaho Space Grant Consortium. I thank Jeppesen and Cessna's Aircraft for allowing me to reproduce copyrighted material. Cessna's material is subject to copyright protection and is not to be further reproduced without the prior written approval of the Cessna Aircraft Company.

I'm not sure which parent interested me more in navigation: my father, a naval officer in World War II, inspired the action part, and my mother, a math teacher, inspired the contemplation. Navigation takes both, so I dedicate this book to both of them.

James S. Wolper
Pocatello, Idaho

Navigation is a science, a craft, and an art. It is a science because one must understand the mathematics of spherical geometry to perform the calculations, and one must understand the shape and the physical properties of the Earth, its atmosphere, and the instruments in order to make the observations. It is a craft because of the judgment that a prudent navigator must employ before risking lives based on his or her solutions. And it is an art because the navigator allows us to see something—our place on the Earth and its relationship with other places—we had not seen before. This book combines all three points of view.

The Global Positioning System (GPS) and other new technologies have made one navigational problem—that of determining present position—easier to solve. But in order to exploit these new technologies, pilots need to understand them. And the new technologies have introduced new problems. Pilots now need to know more about how charts are constructed, about how coordinates of ground positions are determined, and about the nature of the Earth's surface. And the oldest and most important problem, that of developing and exercising navigational judgment, remains.

This book assumes very little mathematical background (all of its mathematics comes from the standard high school curriculum) but takes a simple, original, and rigorous approach to the mathematics of navigation and puts the new technology into historical and practical context. Its treatment of celestial navigation is all new; and, while there are few current airborne practitioners of celestial navigation, modern techniques are based on the same geometry and the same mathematics.

Navigation and Experience

This book began with my father, who was a naval officer in World War II. He has stories. Some are funny: his first night at sea was rough, and he was pitched through a cabin door and landed on top of the sleeping captain. Some are more serious: he sent a convoy to chase a suspicious radar return, which turned out to be a flock of birds. And some are about the practice of navigation: enroute to tiny Eniwetok, they were unable to make celestial sights due to weather. He and another officer decided to try the ship's LORAN.[1] The LORAN worked and gave them some indication of their position, but the captain refused to believe it.

As a new private pilot, I borrowed my father's copy of *The American Practical Navigator*, by Nathaniel Bowditch.[2] The first edition of Bowditch[3] was published in 1802, and it is still in print. Bowditch was America's first mathematician. He was a self-taught former sailor who won international fame for his astronomical studies and his English translation of Laplace's *Traité de mécanique céleste*. His view of navigation was based on experience as well as theory. *The American Practical Navigator* states:

> Until such time as mechanization may become complete and perfect, the *prudent navigator* [my emphasis] will not permit himself to become wholly dependent upon "black boxes" which may fail at crucial moments . . . [t]he wise navigator uses all reliable aids available to him, and seeks to understand their uses and limitations.

The two young officers were right to try to use the new (but unproven) device, and the captain was right to question the results.

Navigation and Mathematics

While I was becoming more adept as a navigator, I naturally became curious about the mathematics of navigation, especially long-range navigation. Long-range navigation is intricate since you can no longer pretend that the Earth is flat. Beginning pilots don't study long-range navigation; airline captains and military pilots learn the subject in special schools.

The most famous tool of long-range navigation is the sextant. A sextant measures an angle, and considering that one of the primary mathematical tools in my Ph.D. dissertation was a very deep generalization of trigonometry,[4] I thought that I knew a lot about angles. But the mathematics of celestial navigation was different in a way that made me uncomfortable: all of the books I could find based their results on "spherical trigonometry," something that I had seen in old books but had never studied. Solid geometry, of which spherical trigonometry is part, had been removed from the standard university mathematics curriculum a little after I was born.

At this point I had been misled enough times by faulty instruments and faulty calculations (even professional mathematicians do this), and I had begun to develop the skeptical attitude that a prudent navigator needs in order to survive.[5] Even one approach to the wrong airport with your father on board makes a convincing case for preflight preparation.

As a prudent navigator, I was surprised that the whole enterprise of long-range navigation was based on a forgotten subject. The laws of spherical trigonometry are correct, but no text that I saw verified them. That meant that celestial navigation had passed from the realm of science into the realm of superstition.

Could this be correct? I flew from my home in rural western Massachusetts to the New York area to examine the New York Public Library's navigation collection. The magic of airplanes: on the same day I could walk both Times Square and a New England village. The library staff patiently brought me every suitable-looking book in the catalog. In the end I was disappointed. Some books told the new navigator to just look it up in the table; some supplemented this with a formula or two, suitable for a hand calculator; some justified the formulas with reference to spherical trigonometry. None proved anything about spherical trigonometry.[6]

This seemed a dangerous attitude in a practice that can kill, and as a mathematician/navigator, I wanted to improve the situation. I had seen some spherical geometry calculations done with vector methods, and this felt like a good approach. Everyone in science or engineering studies vector algebra, and it is easy to prove its correctness. I was fond of teaching that "the best method humanity has for measuring angles is the dot product," so I systematically exploited this idea to rederive all of the formulas of celestial navigation. I wrote a short article about this and used it as the basis of a general audience mathematics course at Mount Holyoke College.

I suppose I could have used the vector methods to rederive the laws of spherical trigonometry and thus allow the practice of navigation to continue as it had been. This didn't occur to me at the time, and now I am glad that I didn't do it, because something as critical as navigation should be based on simple, well-understood mathematics.

And there's a bonus: since that time, satellite-based navigation systems (the United States' Global Position System and the Russian GLONASS) have flourished and have made fundamental changes in the way geographers, surveyors, and navigators measure and model the Earth. The basic operation of these black boxes in determining position is different from celestial navigation, but once the position is determined, the black box uses the same mathematics as celestial navigation uses to work its magic. You may think that the sextant will become a thing of the past, but its mathematics will live on.

Besides, many do not think that the sextant will become a thing of the past. As recently as 1996, the U.S. Navy practiced celestial navigation every day on every ship.[7] Navies the world over are proud of their traditions, but they are also practical. Celestial navigation is inexpensive;[8] no enemy can interfere with celestial navigation; and celestial navigation provides a means of showing new navigators the vagaries of observation and measurement, and thus helps them develop the proper prudent attitude. Finally, celestial navigation is based on measurements that a person can make; this can be a great source of comfort when the chips[9] are down.[10]

Who Should Read This Book

This book is written for both my flying students and my mathematics students. It is not a textbook on advanced mathematics. Every bit of mathematics used is developed from the Pythagorean formula and simple properties of circles and triangles; a high school student can read it. My flying students sometimes have trouble with navigation because their mathematical background isn't deep enough, and my mathematics students sometimes have trouble because their textbooks don't show them how the material is used. This book addresses both issues.

But it also addresses the development of the prudent navigator and so is suited for more advanced navigators. Navigation needs both action and contemplation, and the navigator who neglects the calculations may get into trouble. Here's an example. One of my flying students was having trouble with a calculation and asked me to fly while she worked on the problem. I had asked her to plan a diversion to a different airport; such exercises are important in pilot training because they are common in actual flying, whether in small planes or large jets. I agreed to fly while she worked, and told her that I would continue on the same heading until she told me otherwise. I knew that we were pointed at a mountain 15 miles ahead; she forgot. When she looked up from her calculations, we were barely 500 yards from the mountain. She needed common sense as well as more practice doing the problem.

What You'll Learn

This book begins by investigating the shape of the Earth. Eratosthenes, assuming a spherical Earth, was able to estimate its radius. Understanding the geometry of that calculation leads us through coordinates in the plane and on the sphere, the basic definitions of the trigonometric functions, and basic facts about the geometry of spheres and circles.

I would wager that most of what you "know" about the Earth is based on assumptions, and you will see how assumptions can lead you astray. You'll also see that assumptions are necessary. This also forces us to discuss error, an important problem in navigation. There are two kinds of error: errors of assumption or method, and blunders.

Since the Earth is *not* a perfect sphere, its actual shape is discussed and modeled, and the problems of measuring latitude (there are three kinds of latitude) are discussed. Longitude is more difficult in practice, while latitude is more difficult in theory.

Chapter 2 contains all of the mathematics needed, in the form of vector algebra. This is developed from scratch. Its point of view is a little different from the standard presentations, so those who know something about vectors will find it useful to read. I have used standard notation, so those who can't wait to skip ahead will know what the rest of the book is saying.

The third chapter is about celestial navigation. I have included some new material on azimuth sights. Azimuth sights are less accurate than other sights, but every aircraft is equipped with a compass. The mathematics of celestial navigation is developed completely. The mathematical techniques of a great-circle route are described. There are also some useful approximate techniques, including a section on making your own Mercator chart.

The fourth chapter is about satellite navigation, mostly GPS. Satellite navigation is wonderful but it, too, is subject to errors and blunders. In particular, effective satellite navigation depends on having a model of the Earth's shape. I have not seen this discussed in any of the books on navigating by GPS.

The fifth chapter is about how to navigate in aircraft. It is written with my flying students in mind, but the same ideas apply to all forms of navigation, so it is of interest to sailors and hikers, too. There are no instructions on how to use particular makes and models of the various black boxes; it's more important to concentrate on the general principles that make them work. Both analytic and iterative techniques are illustrated: the former lead to greater understanding, but the latter lead to usable numbers.

The sixth chapter shows how to plan long-range flights over remote areas. There are a lot of calculations, but no new mathematics.

There is a guide to further reading and an extensive bibliography on navigation.

Throughout the book I have included historical notes and anecdotes about navigation, because navigators must develop judgment as well as the ability to calculate.

Mathematics, Again

One warning: some of the methods developed here are the most elegant and appealing to a mathematician. They are the methods that are easiest to understand, not necessarily easiest to use. The new or experienced navigator needs depth of understanding in order to recognize that the black box is about to do something dangerous. The methods here are correct, but they are not always the easiest to use in practice. The practice of navigation is made slightly more complex but infinitely more reliable by preprinted forms, preprogrammed electronic calculators, personal computers, and the like. Someone who understands what is done here can easily learn to use these other tools, and someone who really wants to navigate a ship or airplane *should* use these tools. Someone who doesn't understand can enter the wrong data into a calculator, radio, or computer (have you ever dialed a wrong number?) and will not recognize that the solution proposed is incorrect.

Nor does the book contain every method of navigation ever invented. An example of a missing topic is how computers are able to make landings (Autoland); its treatment requires too much advanced mathematics.

It's important for both the pilots and the mathematicians that the last two parts of this book are about aeronautical practice. Beware: no book can re-create the feelings that come from having the world, which doesn't look like what you had imagined from the map, rush by faster than you had imagined it ever possibly could. In real life, checkpoints are obscured by clouds or fog, radio stations go off the air for maintenance, charted railroad tracks are torn out, low-frequency signals are altered by thunderstorms, LORAN coverage is spotty, GPS satellites go off the air for orbital repositioning, the course you've plotted is unusable due to a forest fire or thunderstorm, and bulbs burn out in airport beacons.[11] Through this all, you have to keep track of where you are and where you want to be.

If you are not a pilot, you can get the whole navigation experience by finding a friend with an airplane or a boat, and trying it! Get a chart, plot bearings, listen to radio static, and the like. Failing that, buy a topographic map of your area and a compass, and learn how to use them. You'll find that action and contemplation complement each other, and that neither will suffice.

This book sometimes refers to a "pilot" and sometimes to a "navigator." In modern practice, this is usually the same person; in an aircraft with a crew of two or more, these duties are shared. This book is aimed at the navigator; it won't tell you how to make better landings, but it will help you make sure you land at the correct airport.

A Note on Reading Mathematics

Navigation is a combination of action and contemplation, and most of the contemplation involves calculations. It's often better to skim each calculation and note its main features before plunging into a symbol-by-symbol analysis. Approach a complex calculation like a fine meal: first note the composition and general external appearance and fix the goal firmly in mind before beginning the detailed dissection.

And remember: a statement you find puzzling is often made clear by a statement on the next page. So keep going. When the thinking gets tough, go out and fly somewhere, and when you get back you'll have more to think about.

NOTES

1. Long Range Aid to Navigation, a then new electronic device.
2. This is in the bibliography as (Bowditch).
3. Nobody calls it by its title.
4. Theta functions.
5. I'm sure that this skepticism was part of the captain's rejection of the LORAN information.
6. I found one notable exception years later: the British Admiralty's *Admiralty Navigation Manual* does include an appendix in which the basic facts about spherical trigonometry are proven.
7. See (Kaplan). Since that article was published, the Navy has announced that it will no longer teach celestial navigation at the Naval Academy, but naval officers have assured me that it will still be done at sea.
8. See Chapter 4 for a discussion of the real cost of GPS, which involves much more than the purchase price of the receivers.
9. Pun intended.
10. See (Callahan) for the experience of improvised navigation. The author, adrift in a lifeboat, made crude navigational instruments and was able to do some navigation using them.
11. One aspect in the unfortunate crash of the USAF transport carrying Secretary of Commerce Ron Brown was that the more precise navigational system that once had been installed at Dubrovnik had been *stolen* during the war in the former Yugoslavia.

CHAPTER 1

The Shape of the Earth

Navigation involves moving along or above the surface of the Earth (or other planet), and in order to understand navigation we need to understand the Earth's shape. Early pilots followed roads or railways and could ignore this, and many pilots fly this way today. But modern aviation is a global enterprise, and with the introduction of the Global Positioning System (GPS) modern pilots need to understand how charts are constructed, how coordinates of ground positions are determined, and more about the nature of the Earth's surface.[1]

Why do we need to know this? As you probably know, we describe a position on the Earth's surface using latitude and longitude, and we also use these coordinates to describe the places we want to go. Since we want to know how far and which way to go, we (or our computers) need to understand the geometry of the planet. And to know where we are, we typically measure the distance and direction we have gone, so we need to know how to interpret these measurements to get latitude and longitude. Thus, a detailed model of the Earth's surface is a basic need for the practice of navigation.

In this chapter we will study the Earth's shape, how the shape is known, how the size of the Earth is known, what this means about charts, the meaning of latitude and longitude, and the importance of knowing the actual (rather than theoretical) shape of the Earth. In order to do this we will review some simple mathematical tools that will also be helpful in later work.

One theme of this book is that navigation is a combination of action and contemplation. You need to do some calculations (the contemplation part), but then you need to go out, get moving, and see where you end up (the action part). A prudent navigator strives to perfect both.

The Earth Is Round

Virtually everyone believes that the Earth is "round." In the twenty-first century, we have been lucky enough to see photographs taken from space, and the regular use of artificial satellites for communications, geodesy, surveillance, remote sensing, and navigation confirms that the round Earth model works, at least to a certain level of accuracy.

In the United States, this idea is popularly attributed to Columbus, but the idea of a spherical Earth is much older. Aristotle reasoned that the Earth was round based on common observation: a departing ship's hull disappears over the horizon before its masts, and this happens in every direction (Wilford). The stars change as one moves north and south. The shadow of the Earth cast during a lunar eclipse was further evidence.[2]

The Greeks liked to measure, and given the model of the Earth as a sphere, it was natural to determine its size. A sphere has one size parameter, its *radius*. Another size parameter is the *circumference*. If you know one of these, then you know the other, since the circumference of a circle is 2π times the radius, where π is about 3.14159... or 22/7. So to measure the Earth, you need to measure its radius. How is it possible to measure such a vast distance in space?

Knowing the circumference of the Earth means that we know how much range to build into our airplanes. It means that we can make inertial navigation (see Chapter 5) work reliably, and it means that we can interpret the signals from GPS satellites accurately.

Eratosthenes, a Greek polymath, gave the first known estimate of the circumference (and hence the radius) of the Earth. In general,

Greek scientists like Eratosthenes used clever abstract reasoning like the deductive geometry of Euclid to determine mathematical properties. Euclid's geometry was based on *axioms*, such as "through any two points pass exactly one line." But we will use other tools, namely, *coordinates*, and problems will be easier to solve if we use them than if we limit ourselves to the classical tools of unmarked straightedge and dividers.

Eratosthenes had heard that in Syene (modern Aswan) the Sun shone straight down a well at the solstice. A solstice occurs twice each year when the Sun's apparent position reaches its northernmost and southernmost points. This happens around June 21 and December 21. Another way of thinking about this observation is that a stick pointing straight up would have no shadow at that instant. On the day of the solstice, Eratosthenes measured the length of a stick's shadow in Alexandria; he estimated the distance to Syene; and thus he was able to estimate the Earth's circumference. He used formal geometric reasoning, but it's also possible to understand this calculation using the more modern method of vectors.

Vectors and Coordinates

Navigators have always used vectors in the style of physics, that is, as entities with direction and magnitude. The wind is an important example. We will use vectors in a more mathematical way based on coordinates. GPS uses these coordinate methods in its calculations. To begin, we will review coordinate geometry.

First, imagine moving on a *line*: we have only one direction of movement (see Figure 1-1). To put coordinates on the line, we choose a starting point (called the origin, and often denoted *O*), and we measure off one positive unit of distance. If the positive unit is to the right, then the coordinates of points to the left of *O* have negative values.

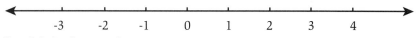

Figure 1-1. Coordinates on a line.

Now imagine moving on a *plane*: There are evidently an infinite number of directions of possible movement, but we can single out two of them and determine every direction in terms of those two. We still have to pick an origin O, and a unit distance. But now we can go to the right from O (you can imagine a horizontal line drawn through O), and we can also go up from O (imagine a vertical line). In this coordinate system, left and down are negative (see Figure 1-2).

To reach any point on the line, we specify the number of (positive or negative) units to move from O, and on the plane we specify the number of left–right units and the number of up–down units. Thus the line has one dimension, and the plane has two. We imagine in both cases that we can go an unlimited distance in any direction; this may not be practical, but it has the philosophical advantage of pushing some of the difficulties out of sight.

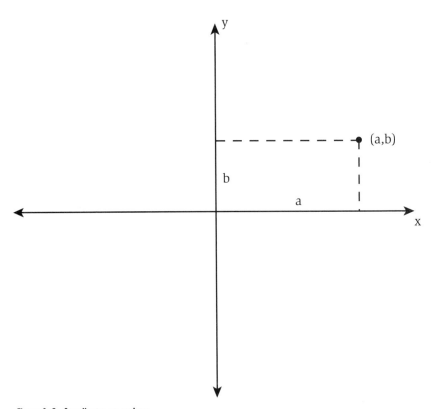

Figure 1-2. Coordinates on a plane.

A point on a line can be represented by a number: "14" means the point that is 14 units to the right of O, and "$-\frac{3}{2}$" represents the point that is $\frac{3}{2}$ units to the left of O. It takes two numbers to represent a point on the two-dimensional plane, and we make the arbitrary convention that the first number indicates left–right movement and the second indicates up or down. Thus, $(14, -\frac{3}{2})$ is the point 14 units to the right from O and $\frac{3}{2}$ units down.

What happens in 3-space? Look at Figure 1-3. In a 3-dimensional space we can move left or right, forward or back, or up or down, so there are 3 coordinates and therefore 3 dimensions. The convention is that the first is left–right, the second forward–back, and the last up–down. Thus, $(14, -\frac{3}{2}, 11)$ means the unique point obtained by starting at O, moving 14 units to the right, $-\frac{3}{2}$ units forward (that is, $\frac{3}{2}$ units back), and 11 units up.

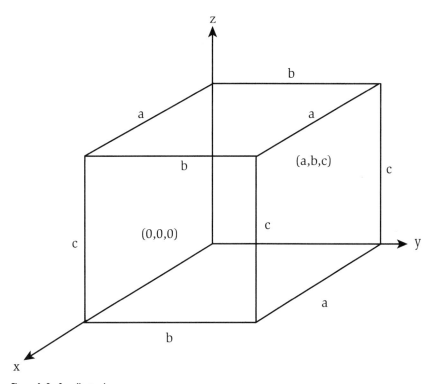

Figure 1-3. Coordinates in space.

We specified the coordinates in a certain *order*; mathematicians call this *orientation*. The standard in a plane is to rotate *counterclockwise* from the first to the second. See Figure 1-4.

In space, we use the *right-hand rule*: the first coordinate is aligned with the first finger of the right hand, the second is aligned with the second finger, and the third is parallel to the thumb.

Lines, planes, and spaces do not come equipped with coordinates. It is up to us to determine where O is, and to determine in which direction the coordinates' axes point. This is useful. For example, we use coordinates to describe points on the Earth's surface, and this is typically done by saying "Let O be the center of the Earth, let the third axis point through the North Pole, and let the first axis point through the prime meridian."[3] The direction of the second axis is, of course, determined by the right-hand rule.

Some Remarks on Variables

We can't do anything with coordinates unless we allow them to vary, and that means that we have to use letters rather than numbers to represent the various quantities. There are some conventions in this regard, but there are no hard-and-fast rules, so you have to pay attention when someone says what a variable stands for. Usually, a variable point in space is denoted (x, y, z), but sometimes x has another meaning, and there is no reason for us to be locked in to these letters. Thus, sometimes we refer to a point as (a, b, c) or

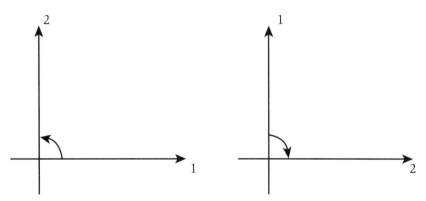

Figure 1-4. Two orientations of the plane.

(x_1, x_2, x_3) or even $(x, y, 5)$ (which occurs if we happen to know that the last coordinate is 5). In mathematics, as in any careful form of rhetoric, we define our terms. If we have two points (x, y_1) and (x, y_2) in the plane, then we know that they have the same first coordinate. The letter isn't important; the relationship ("same" or "different") is crucial. On the other hand, if we have no reason to believe that two points are the same, then we had better make sure that the variables describing their coordinates are different; we write (x_1, y_1) and (x_2, y_2), letting the subscript remind us that x_1 is related to but not necessarily equal to x_2.

The Size of the Earth: Distances in the Plane and in Space

Eratosthenes was interested in measuring a distance: the radius of a sphere is the distance from its center to any of the points on its surface. It doesn't matter which surface point we choose, because the definition of a sphere guarantees that all points are at the same distance from the center.

   Along a line, each point is represented by a single coordinate. Let's suppose that we have points at a and b. The distance is related to the difference between these numbers; that is, the distance is either $a - b$ or $b - a$. Since these are variables, we don't know which one is bigger. The distance has to be positive, but if we subtract the larger from the smaller (by accident), the number we get won't be valid. Thus, we make an arbitrary choice, and use the *absolute value* to make sure that what we get is positive. Algebraically, $|x|$ depends on whether x is positive or negative ($|0| = 0$). If x is positive or zero, then $|x| = x$; if x is negative, then $|x| = -x$. This can be confusing, but $|-7| = 7 = -(-7)$. Geometrically, $|a - b|$ is the *distance* between a and b.

   In order to use coordinates to make the computation of distance in the plane a lot easier than it was for Eratosthenes, we will use the *Pythagorean formula*, which relates the lengths of the three sides of a right triangle.[4] As Figure 1-5 shows, if the two sides that meet at the right angle have lengths a and b, and if the hypotenuse has length h, then

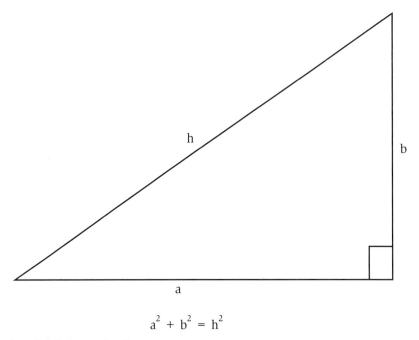

$$a^2 + b^2 = h^2$$

Figure I-5. Pythagorean formula.

$$a^2 + b^2 = h^2.$$

Let's put this another way. Suppose that the point opposite the side with length a is called A, the point opposite b is called B, and the point opposite h is called H. Point H is where the right angle is.

Now the length h is the *distance* from point A to point B. Thus, the Pythagorean formula has direct application to our problem of finding distance. Let's go back to the coordinate plane, and suppose that point A is the higher of the two points and that it has coordinates (a_1, a_2), and that point B has coordinates (b_1, b_2). (If the two points are at the same height, we can either measure along the horizontal or arbitrarily pick one as A.) Then the coordinates of H are (a_1, b_2): the first coordinate is the same as A's because H is directly under A, while the second is the same as B's because H is at the same height as B.

Using what we know about measuring along a line, then, the length of the vertical leg, b, is $|b_2 - a_2|$, and the length of the horizontal leg, a, is $|b_1 - a_1|$. Thus, the length of the hypotenuse—the distance from A to B—is $\sqrt{|b_2 - a_2|^2 + |b_1 - a_1|^2}$. Since a square is never negative, the absolute values are not needed, and we can simplify the distance to $\sqrt{(b_2 - a_2)^2 + (b_1 - a_1)^2}$. This is the *distance formula* in the plane (see Figure 1-6). For example, the distance between (3, −1) and (−2, −4) is

$$\sqrt{(-1-(-4))^2 + (3-(-2))^2} = \sqrt{3^2 + 5^2}$$
$$= \sqrt{9 + 25}$$
$$= \sqrt{34}$$
$$\approx 5.8.$$

(One of Pythagoras' disappointments was that square roots, which are so necessary for doing actual measurements, are generally irrational numbers, so we have to approximate them. The number of dec-

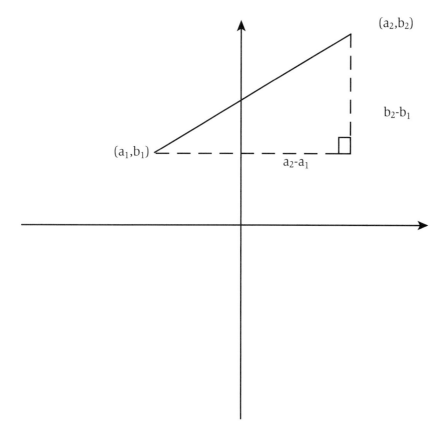

Figure **1-6.** Calculating distance in the plane.

imal places to carry depends on the situation. We will discuss
Newton's method for estimating square roots in Chapter 5.)

What about distance in space? The geometric problem we need to
solve is that of finding the length of the diagonal of a box, as in Figure
1-7.

Since all of the corners of a box are right angles, we can use the
Pythagorean formula with the triangles ABC and ACD. For the first,
AC is the hypotenuse, and its length is $\sqrt{x^2 + y^2}$. For the second, the
two sides have length $\sqrt{x^2 + y^2}$ and z, so the diagonal has length

$$\sqrt{(\sqrt{x^2 + y^2})^2 + z^2} = \sqrt{x^2 + y^2 + z^2}.$$

Now we can determine *equations* for circles and spheres. A word
about equations: you have to be aware of what the variables mean,
but the most important part of an equation is the *relationship*
between the variables. When we write $y = x^2$ as an equation in the
(x, y)-plane, all we are really saying is that the second variable is the
square of the first. So, if the first variable is k and the second is z, the
equation changes to $z = k^2$. The problem is that a wordy formulation

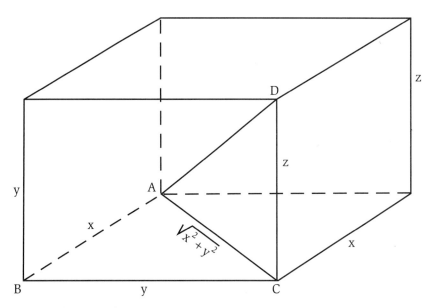

Figure 1-7. The diagonal of a box.

is hard to work with, so we use the single letter x for "the first variable" and the single letter y for "the second variable."

An equation for a curve in the plane defines a relationship between the coordinates of the points on that curve. Every point on the curve must satisfy this relationship, and no points not on the curve should do so. A similar idea holds for a surface in space. When we talk about a curve in space, though, there is a big difference. A relationship between the coordinates means, roughly, that we can determine one of the coordinates if we know the rest. In the plane, an equation enables us to determine x (or something about x) if we know y, or to determine y (or something about y) if we know x. But for a curve in space, an equation determines the third variable from the other *two* variables, so to describe a curve in space we need at least two relationships.

What is a circle? The definition is "the set of points in a plane at a fixed distance from the center." This is hard to work with, so let's use coordinates. First, let r denote the radius. In some senses, r is fixed, since for any one circle that we have in mind, the radius is fixed, but there is a sense in which r is a variable (for example, when we are talking about *all* circles). Then, let the center be at the point (h, k).[5]

Now, let (x, y) be a variable point on the circle. That is, we allow (x, y) to denote any point on the circle, without being more specific. Then the definition says that the distance from (x, y) to (h, k) is r. We translate this into symbols using the distance formula, and simplify:[6]

$$\sqrt{(x - h)^2 + (y - k)^2} = r \quad \text{so}$$
$$(x - h)^2 + (y - k)^2 = r^2.$$

The same reasoning works in space. If the center is (h, k, j) and the radius is r, then the equation of the sphere with center (h, k, j) and radius r is

$$(x - h)^2 + (y - k)^2 + (z - j)^2 = r^2.$$

The Size of the Earth: Distances on the Sphere

For the moment, we will assume that the Earth is a perfect sphere; we'll find out more about the actual shape later. Eratosthenes needed to measure the distance from Alexandria (where he lived) to Syene along the surface of the Earth. This is not a straight line, so he had to use a measurement along a circle.

Speaking loosely, Eratosthenes paced the distance out (that is, he took the reports of caravan members). It appears that we could do the same thing by using a piece of string to measure distances along the surface of a globe, but this would be taking advantage of the fact that someone has already measured that distance and scaled it down to your globe's size.

The pacing method can't be used for every measurement. For example, it's not possible to pace the distance to the Moon. This would seem at first to be a disadvantage, but we actually gain something because we will have to develop other ways of *inferring* distances. Some of these methods are easier and more accurate, too.

In the case of the Moon, the first estimate of its distance was a consequence of multiple simultaneous observations of Venus crossing the face of the Sun; this is called a *transit*. The idea was Halley's, and one of the observers was taken to Tahiti by Captain James Cook, one of history's greatest navigators. We will discuss this later.

To measure distances along a sphere such as the Earth, it is enough to know how to measure along a circle. Why? Imagine points A and B on a sphere centered at the origin O. The three points O, A, and B define a plane, and we can choose (x, y, z) coordinates in space so that this is the plane where $z = 0$, while the sphere has equation $x^2 + y^2 + z^2 = r^2$ (r is the radius). The points A and B satisfy both equations, so, substituting $z = 0$, we find that both A and B satisfy $x^2 + y^2 = r^2$; in other words, they lie on a circle in the plane $z = 0$. See Figure 1-8.

It is intuitively clear that the distance from A to B is the length of this circular arc, and empirical evidence for this comes from Kepler's laws about planetary motion. Since a natural or artificial satellite "naturally" travels a minimum distance, and since, per Kepler, it travels in a plane through the center of the planet, the minimum distance along the planet's surface must be where such a plane hits the surface. This is the definition of a great circle.

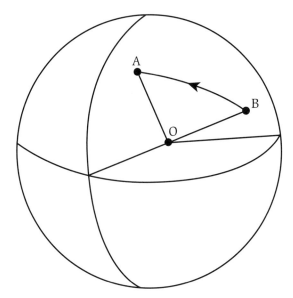

Figure 1-8. Distances on a sphere.

Why is this evidence empirical? Since Newton, most presentations of Kepler's laws are based on calculus, but Kepler based them on the astronomical observations of Tycho Brahe, who had spent many years making painstaking measurements. Ironically, Kepler made many errors in his derivations that cancelled each other out, so he arrived at correct conclusions. See (Bate, Mueller, and White).

Special Case: Distance on a Circular Arc

Now, back to distances along circles. Recall that a circle with radius r has circumference $2\pi r$, where π is approximately 3.14159, or about 22/7. The distance along the circle between two diametrically opposed points is obviously half of the circumference, or πr. See Figure 1-9.

If the points A and B in Figure 1-9 are separated by a right angle, then the distance between them is one-quarter of the circumference: $2\pi r/4$ or $\pi r/2$. In general, if the smaller "piece of pie" defined by the two points A and B is some fraction $1/f$ of the circle, then the distance between A and B is $2\pi r/f$.

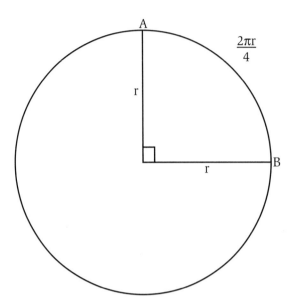

Figure 1-9. Distances along a circle.

The two radii—the lines from the center to *A* and from the center to *B*—meet at an angle, which we can call *a*. If we know the angle *a*, we can draw the points and measure the distance; and if we know the distance, we can measure the angle. On a circle, an angle is a distance, and a distance is an angle. If the distance is $1/f$ times the circumference, then the angle is $1/f$ times . . . what? How do we measure angles?

There are three common measures of angles in use today: *radians*, *degrees*, and *grads*. The last two are the easiest to describe: the angular distance all the way around a circle is 360 degrees or 400 grads. You measure these with a protractor, which works for a circle drawn on paper but cannot be used at the center of the Earth.

The use of a 360-degree circle in navigation is fairly new. Mariners used to divide the circle of the compass into 32 *points*. Four of these were the cardinal headings North, South, East, and West. Halfway between these were four more: Northeast, Southeast, Southwest, and Northwest. These were further divided into North-Northeast, East-Northeast, etc. Dividing each of these in half led to a total of thirty-two. Each point was a little more than 11°. In many ways, this was more than enough accuracy in measuring the angle. Typical practice at sea was to sail off at an angle until reaching the proper latitude, then to sail due east or west.

You measure radians with a string. Since an angle at the center of a circle corresponds to a distance along the circumference, you construct the angle you want to measure at the center and measure the corresponding arc along the circle. But, the measure you get depends on the radius of the circle, so you divide by the radius. See Figure 1-10. Thus, 2π radians is $360°$ or 400 grads, so each radian is $\pi/180$ degrees or $\pi/200$ grads.

Which is better? It's a general rule in mathematics that if we have developed two ways of looking at something, then at various times we need each of them. Neither is better by itself; given special circumstances, one may be the best. For calculus radians are best, but for astronomy and navigation degrees work pretty well. One reason is that the number of degrees in a whole circle is very close to being the number of days in a year—and this is no coincidence. Another reason is that 360 is a very easy number to work with, since it is evenly divisible by 2, 3, 4, 5, 6, 8, 9, 10, 12, 15, 18, 20, 24, and many more numbers. Thus, 8 degrees is $1/45$ of a circle.

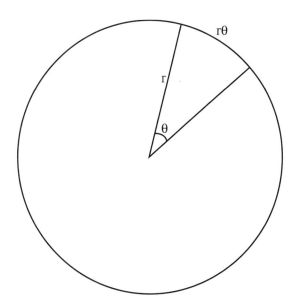

Figure 1-10. Radian measure.

The degree is itself divided into 60 pieces, called *minutes*, each of which is divided into 60 *seconds*. A half of a degree is 30 minutes, written 30′. One quarter of that is 7.5′ or 7′30″. All the way around a circle is 360 × 60 = 21,600 minutes.

Two common units of length are closely related to these ideas: the meter and the nautical mile. Both were designed to make it convenient to measure along the surface of the Earth. The *meter* is a little more straightforward because, like everything else in the metric system, it was designed by dividing something by a power of 10. The original definition was that there should be 10,000,000 meters from the equator to the North Pole *along the meridian of Paris*. Since a kilometer is 1000 meters, and since the distance from the equator to the North Pole is one-quarter of the circle, this means that the metric circumference of the Earth was taken to be 40,000 kilometers. (Later refinements of the meter and improvements in understanding the shape of the Earth changed this, but 40,000 km is a good top-of-the-head estimate.)

All the way around a circle is 21,600 minutes, and this leads to the definition of the *nautical mile* as the length of 1 minute of latitude.

So, a nautical mile was originally 40000/21600 or approximately 1.85185 kilometers. The modern definition is 1852.0 meters exactly, and the circumference of the Earth is taken to be 21,602.518 NM, or 40,007.863 km.

The statute mile of 5280 feet doesn't fit any of these schemes. A nautical mile is 6076.2 feet or about 1.15 statute miles.

The Size of the Earth: Circles and Tangents

Eratosthenes deduced the circumference of the Earth using some clever geometry and radian measure. As mentioned earlier, he lived in Alexandria near the mouth of the Nile at the time, and he had heard that the Sun shone straight down a well at the solstice in Syene (modern Aswan). Equivalently, a stick pointing straight up would have no shadow at that instant. Remember that Eratosthenes measured the length of a stick's shadow in Alexandria on the day of the solstice; he knew (or thought he knew) the distance to Syene; and thus he was able to deduce the Earth's circumference.

How? The shadow's length allowed Eratosthenes to deduce the angle between the *zenith*—the point in the sky directly overhead—and the Sun's position in the sky. Since at the same instant in Syene the corresponding angle was known to be zero, he could deduce the angle *AOS*, where *O* is the center of the Earth. See Figure 1-11. In other words, Eratosthenes knew the radian measure of *AOS* and the measure of the corresponding arc, so he knew the radius.

This explanation raises more questions than it answers. How could he deduce the angle between the Sun and the zenith from the length of a shadow? What does it mean for a stick to point straight up? How did he know that the distance was used correctly?

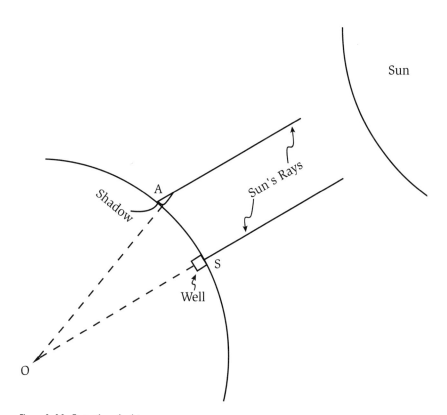

Figure 1-11. Eratosthenes's picture.

Moreover, the result he got depends very much on the *assumption* that the Earth is a sphere. We will reinterpret the calculation based on other assumptions, and the results are equally plausible.

Where Is Straight Up?

What does it mean for a stick to point "straight up" from a curved surface? For a general surface, this question is pretty difficult to answer (you need calculus), but the sphere is a particularly nice surface, and we can answer the question quite precisely. We'll start by considering a circle, and then argue by analogy about spheres.

For any curve C, a line L is *tangent* to C at the point P if L meets C at P and no other nearby point. There is an exception: a line is tangent to itself. See Figure 1-12.

In the special case when the curve is a circle, the tangent line is easy to find. First we'll consider a particular circle as an example. Usually, you can't determine something about only one member of a family and apply the result to every member of a family (that's why there's algebra), but for the case of circles, we can use symmetry and other properties to draw a valid conclusion. In this case, we'll choose coordinates that make the conclusion easier to see.

Consider the circle C given by the equation

$$x^2 + (y-1)^2 = 1$$

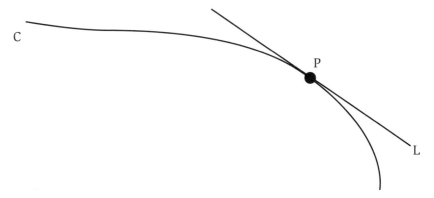

Figure 1-12. Tangent line at P.

using the ordinary coordinates in the (x, y)-plane. The center is at $(0, 1)$, and the radius is 1. This looks like a coin resting on its edge, as shown in Figure 1-13.

It's clear that the circle touches the x axis exactly once, at the point $(0, 0)$, so the x axis (the line with equation $y = 0$) is tangent to the circle at this point. This is verified analytically by solving for the point of intersection: the x axis has the equation $y = 0$, so the intersection is where $x^2 + (0 - 1)^2 = 1$, or $x^2 = 0$. The equation $x^2 = 0$ should have two roots, just as the equation $x^2 = 4$ has the two roots 2 and -2, and the equation $x^2 = 100$ has the two roots 10 and -10. So $x^2 = 0$ must have the roots 0 and -0, which is of course 0. The double root is an indication of a tangency.[7] Normally, a line hits a circle in two distinct points; when the points come together, as they do here, the line is a tangent line.

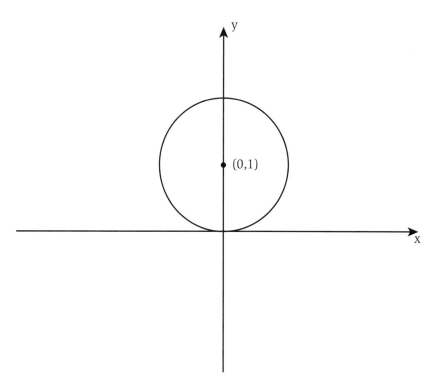

Figure 1-13. Graph of $x^2 + (y - 1)^2 = 1$.

Now notice that the *radial line* through $(0,0)$—that is, the line from the center of the circle to the point—is the line $x = 0$, which is perpendicular to the tangent line. It is clear, using symmetry, that this relationship would not change if we were to rotate the circle (that is, we are using the symmetry of the circle), nor would it change if we changed the radius (we could do this by considering the circle with equation $x^2 + (y - r)^2 = r^2$), nor would it change if we moved the center of the circle elsewhere. In other words: if P is a point on a circle with center O, then the tangent line at P is perpendicular to the line segment $\overline{OP}$.

Analogy with Tangents to Spheres

But we want to work with spheres. A sphere is a *surface*, and so it has a *tangent plane*. For any surface S, a plane L is tangent to S at the point P if L meets S at P and no other nearby point.[8]

As for the curve case, the general solution to the problem of finding a tangent plane to a surface requires calculus, but for a sphere there is a solution analogous to the solution of the tangent problem for a circle. There are several approaches. One is to replace the "coin on its edge" analogy with the "basketball on a floor" analogy: the floor is the tangent plane. Using equations, the sphere

$$x^2 + y^2 + (z - r)^2 = r^2$$

which has radius r, meets the plane $z = 0$ at the point $(0,0,0)$ and no other, so that plane is the tangent. Or we can consider great circles passing through the point P of interest; each of these has a tangent *line* perpendicular to the radius, and each such line lies in the tangent plane, so the tangent plane is perpendicular to the radius.

Working with Planes

In fact, the easiest way to work with a plane is to have a line perpendicular to the plane; this is called a *normal* to the plane.[9] See

Figure 1-14. Later on, we will see how a normal to a plane helps determine an equation for the plane.

But we don't need equations now. Look at Figure 1-14: The normal line points "straight up" from the plane, so this is the line we are looking for. "Straight up" from a surface means along the normal line to the tangent plane to the surface.

If we assume momentarily that the Earth is perfectly round, we can determine this direction pretty easily using either a plumb line or a level, just as you would do in setting a fencepost. This is because on a perfect sphere, gravity pulls objects toward the center of the sphere. But this, too, can be deceptive, as we will see later.

Eratosthenes' Normal

So now we can examine Eratosthenes' measurement: His stick (the *gnomon*) was along the normal line to the tangent plane, and this is exactly the line through the center of the Earth. The shadow was measured on the tangent plane, as in Figure 1-15.

Eratosthenes thought that the angle was about 7.2°, so the distance from Alexandria to Syene was about 7.2/360ths of the circumference of the Earth. Since this is about 2 percent, he multiplied the distance from Alexandria to Syene by 50 to estimate the Earth's circumference.

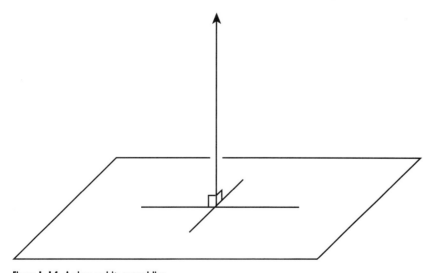

Figure **1-14**. A plane and its normal line.

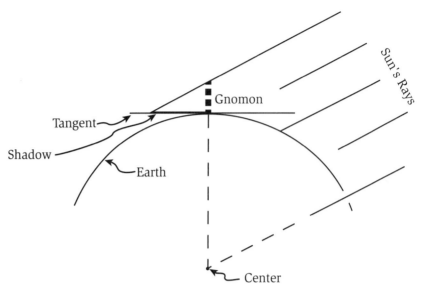

Figure 1-15. The geometry of the gnomon.

The Size of the Earth: Shadows and Measuring Angles

Eratosthenes measured some lengths and obtained angles. If you've seen any trigonometry before (I'm not assuming that you have), this problem should sound familiar.

Trigonometry has a bad name among mathematics students. This is a shame, since it is one of the most useful and interesting subjects known. Trigonometry is also an effective way of working with circles; you might have expected this because the radian measure of an angle is defined using a circle. All of the problems of navigation are solved using trigonometry.

Basic Definitions of Trigonometry

Let's begin with the basic definitions. Consider a right triangle with an angle A as one of the acute angles.[10] The *hypotenuse* is the side opposite the right angle. The other two sides are given names based on their relationship to the angle A: the *adjacent* side touches point A, and the *opposite* side does not. See Figure 1-16.

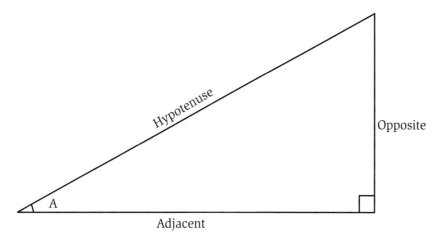

Figure 1-16. Names of the sides of a right triangle.

There are six possible nontrivial ratios that one can make using the sides of this triangle, as shown in Table 1-1.[11] We mostly work with the sine and cosine; as you can see from the table, all of the other functions can be expressed in terms of these two.

Estimation and an Application

One of the reasons people dislike these functions is that they are usually impossible to calculate exactly. For example, my calculator tells me that $\sin 42°$ is 0.669130606, and this is only an approximation. For most practical purposes, four places are enough ($\sin 42°$ is approximately

Table 1-1. Trigonometric Functions

NAME	NOTATION	RATIO	RELATIONS
Sine	sin A	opposite / hypotenuse	
Cosine	cos A	adjacent / hypotenuse	
Tangent	tan A	opposite / adjacent	sin A/cos A
Cosecant	csc A	hypotenuse / opposite	1/sin A
Secant	sec A	hypotenuse / adjacent	1/cos A
Cotangent	cot A	adjacent / opposite	1/tan A

0.6691). On the other hand, there are so many beautiful relationships between these functions that it is often not necessary to actually calculate them. That is to say, in some contexts you can and should look at the expression $\sin 42°$ as a number with a very long name rather than trying to give it another name.

In the practice of navigation, it is often good enough to have a very rough approximation for the values of these functions, and it is possible to do a fair amount of work with them in your head. Here is an important example from aviation. Pilots need to be careful about landing in a crosswind, that is, a wind that is moving across the runway rather than along it.[12] In fact, most airplanes have a crosswind limitation. For training airplanes, it can be as low as 12 knots, while large jets can handle more than 30 knots.[13]

When the wind is directly across the runway, it is easy to determine the crosswind, but what if the wind crosses the runway at an angle? Then the effective crosswind component is the wind speed times the *sine* of the angle between the runway and the wind (see Figure 1-17). So a pilot approaching an airport to land needs to be able to compute sines in order to determine if the airplane can handle the existing crosswind. In fact, most operating handbooks (one of the documents that must be carried in the aircraft) have either a table or graph for this purpose.

Rather than dig out the handbook, a pilot can find an approximate solution by remembering a few numbers and using headwork. Start with these facts:

- If A is in *radians* and $A < 0.5$, then $\sin A \approx A$.
- One radian is $180/\pi$ degrees, which is approximately 60°.
- Sin 45° = $\sqrt{2}/2$, which is approximately 0.7.
- Sin 60° = $\sqrt{3}/2$, which is approximately 0.9.

(The facts about sin 45° and sin 60° follow from elementary geometry.)

These facts, rough as they are, translate into usable information about crosswind components. Let X be the angle between the wind and the runway.

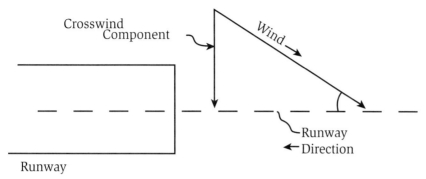

Figure 1-17. Crosswind components.

- If X < 30°, the crosswind component is the windspeed times $X/60$.

- If X is about 45°, use the windspeed times 0.7.

- If X is about 60°, use the windspeed times 0.9.

- If X is more than 60°, use the whole windspeed.

I use this in flight instruction. One blustery day I was with a student working on his commercial pilot certificate. We were returning from a practice session in his airplane, and the wind was from 210° at 25 knots, clearly favoring runway 21.[14] But here was an opportunity; runway 16 was about 45° of the wind, so the crosswind component was about 17 knots since $0.7 \times 25 \approx 17$. This happened to be the airplane's maximum demonstrated crosswind component, so I suggested that he try a landing on runway 16. He hesitated but landed well.

This technique illustrates another important thing about trigonometric functions that seems to give people a creepy feeling: they are nonlinear, in the sense that if you double the angle, you do *not* double its sine.[15]

Relations Between Trigonometric Functions

How are trigonometric functions related to each other? First, remember that all of them can be expressed as functions of sine and cosine alone: $\cot A = \cos A / \sin A$, etc. So, how can we relate the sine and cosine functions to each other?

We do this by looking at the unit circle, whose equation in the usual plane coordinates is $x^2 + y^2 = 1$ (Figure 1-18).

Pick a point P on the circle with coordinates (a, b), and draw the ray that starts at the origin $(0, 0)$ and passes through P. Call the angle between the x axis and this ray A; keep in mind that mathematicians measure angles counterclockwise from the x axis. See Figure 1-19.

We can now construct a right triangle inside the circle by "dropping" a perpendicular from P to the x axis; this hits the axis at the point Q whose coordinates are $(a, 0)$. The triangle we need is OPQ.

The hypotenuse of this triangle has length 1, by construction. Thus, using the definitions above,

$$\cos A \;=\; a/1 \;=\; a \quad \text{and}$$
$$\sin A \;=\; b/1 \;=\; b.$$

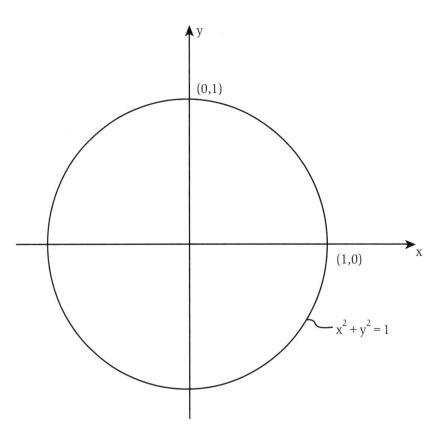

Figure 1-18. The unit circle.

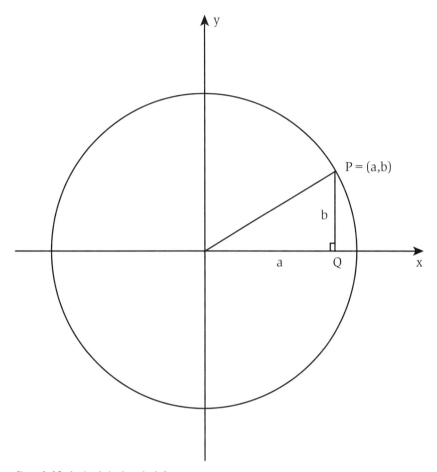

Figure 1-19. A triangle in the unit circle.

Since (a, b) is on the circle, we know that $a^2 + b^2 = 1$, and, reversing the equalities, we see that for *all angles A*

$$\sin^2 A + \cos^2 A = 1.$$

(Notice that we write $\sin^2 A$ rather than $\sin A^2$, which might be ambiguous. What we mean is $(\sin A)^2$, which is awkward.)

Eratosthenes and Trigonometry

We can now see how Eratosthenes computed an angle from lengths. The shadow was the opposite side from the angle he wanted to measure, and the length of the gnomon was the adjacent side. Their ratio gave him the tangent of the angle he needed. By consulting a table of tangents, he could find the angle. According to (Williams), such tables were available at least as early as the second century A.D.

Finding Angles

It's a quirk of vector methods (as you will see in Chapter 2) that they don't give us the angle we want directly; rather, they tend to give us the cosine of the angle we want. You'll see in Chapter 3 that the formulas for distance and bearing actually give us the sine of the distance[16] and the cosine of the bearing. In other words, if we use vector methods to emulate Eratosthenes, we are more likely to end up with the number 0.9921 than with $7.2°$. In fact, 0.9921 is the four-digit approximation of $\cos 7.2°$.

We need a way to use the trigonometric functions backwards. That is, given, say, the sine of an angle, we need to be able to determine that angle. This is an example of an *inverse trigonometric function*.

There's a genuine problem here, since $\sin 32° = \sin 392°$; in fact, there are an infinite number of angles whose sine is the same as that of $32°$. So we'll agree that when we try to find an angle whose sine is 0.7134, we'll look for one between $-90°$ and $90°$. For inverse cosine, we'll pick the unique angle between $0°$ and $180°$ with the given cosine.

Most calculators that do trigonometric functions also do the inverse trigonometric functions. There are two common notations: one can call the angle whose sine is 0.7134 either $\sin^{-1} 0.7134$ (called "sine inverse of 0.7134") or $\arcsin 0.7134$ (called "arcsine of 0.7134"). There are good arguments against both notations. For example, $\sin^2(40°)$ means take the sine of $40°$ (approximately 0.6428) and square it, getting approximately 0.4132). But $\sin^{-1}(.40)$ does *not* mean to take the reciprocal of the sine of $.40°$.

For cosine, we'll agree that the arccosine of some number should be between $0°$ and $180°$. Thus, $\arccos 0.9921 \approx 7.2°$.

One thing to keep in mind: Since the hypotenuse is always the longest side of a right triangle, neither the sine nor the cosine can exceed 1 in absolute value. So, you can't take, say, arccos 1.2. If you find yourself trying to do such an impossible task, then you have made some kind of blunder. You might as well start over.

The Size of the Earth: The Assumptions

It is easy to forget that every measurement involves assumptions, and in some cases it is hard to determine what assumptions are made. In the case of measuring the Earth, there are a lot of assumptions, and navigators who are unaware of them can be fooled; a navigator who is fooled may soon be dead. One example of this is in Chapter 3: Australia used different geographical coordinates than the United States until recently.

The assumptions Eratosthenes made are clear and, in fact, it is quite clear that they are *assumptions*, since there is no way he could have verified them either analytically or experimentally. What assumptions did he make? First, he assumed that the Earth was spherical. He also assumed that the Sun's rays are parallel. These were the major assumptions; there were minor ones as well, such as that a plumb line points straight down and that Alexandria was due north from Syene.

The last of these is the easiest to dismiss: it's wrong. Alexandria is also a little west of Syene, and this introduced an error into the calculation.[17] It did not, however, challenge the whole method; the same method with a correct estimate of the change in latitude would have yielded a good estimate.

The assumption that a plumb line points straight down seems very plausible, but in fact it is false because the Earth (as was discovered in the eighteenth century) is not a homogeneous mass. Some places are denser than others; some are "frothier" than others; and a plumb line on the boundary of two such regions would be more attracted to the denser area and thus would be deflected from the true vertical.

Again, this assumption introduces error into the calculation but it does not invalidate Eratosthenes' method. In fact, the deflection of the plumb line from the vertical is very difficult to measure, and for many purposes (including virtually all construction) this error is safely ignored.

How Close Is the Sun?

The other assumptions are more troublesome and are much harder to dismiss. The plumb line and relative position errors might change the result by a few percent, but the others could lead to complete nonsense.

First, let us look at the assumption that the light rays from the Sun are parallel. We have been told that the distance to the Sun is about 94 million statute miles. We know this because of Cook's second Pacific voyage, one of whose purposes was to observe a transit of Venus.[18] Transits of Venus occur on a periodic basis, but the behavior seems bizarre: transits occurred in 1518, 1526, 1631, 1639, 1761, and 1769. This pattern is then repeated.

The idea in observing the transit was that if it were observed at different places on the Earth's surface, the difference in the observations would, with some trigonometry, determine the distance to the Sun.[19] See Figure 1-20.

The transit observations indicated a parallax of 8.9″. Since the observing locations were separated by about 4000 miles, and the tangent of 8.9″ is about 4000 divided by the distance, the distance to the Sun is about 94,000,000 miles.

Cook was one of the greatest navigators, and like Bowditch was recognized for his contributions to science. These included observations of the solar eclipse of August 5, 1866, his discovery that vitamin C would prevent scurvy, and his explorations of the coasts of Australia and New Zealand. See (Sherry). In his honor, I named my computer after his ship the *Endeavour*.

How Big Is the Sun?

Once we know the distance to the Sun, we can estimate its diameter and conclude that the Sun is much larger than the Earth. But it is not possible to sit in one's study and determine these facts. Cook's voy-

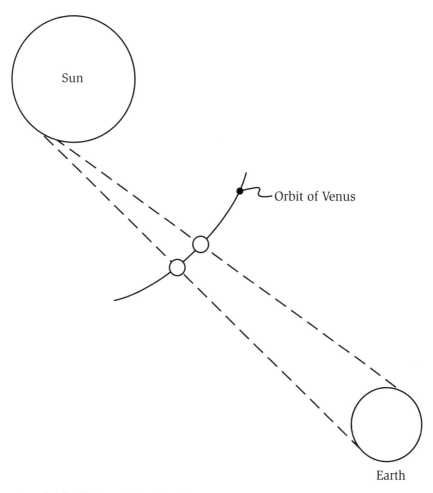

Figure 1-20. Simplified geometry of the transit.

age was no mere exercise in contemplation. So Eratosthenes must have, at least implicitly, taken this assumption on faith.

What if Eratosthenes instead assumed that the Sun is the size of a flea? The picture becomes that from Figure 1-21. The rays are not parallel. But we can still make Eratosthenes' measurements and perform the same calculations.

In this situation, the stick casts a longer shadow; so the tangent of the angle would be too large; so the estimate of the angle would be too large; so his estimate of the Earth's size would be much too small.

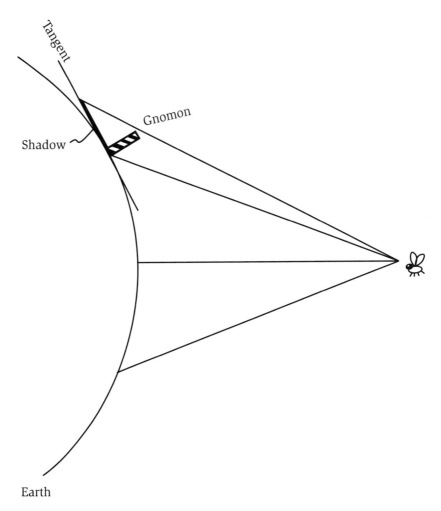

Figure 1-21. Rays from a flea-sized Sun.

What if the Earth Were Flat?

The other assumption is that the Earth is spherical. What if the Earth were indeed flat? In fact, according to (Li and Du), the Chinese scholar Zhoubi Suanjing used a measurement similar to Eratosthenes', but assumed a flat Earth. He used the Pythagorean theorem to determine the distance to the Sun.

We can do this calculation and *incorrectly* determine the distance to the Sun. The distance from Alexandria to Syene is approximately 568 miles, so we have the triangle shown in Figure 1-22 (D is the "distance" to the "Sun").

We would still measure the sun's angle of elevation as $\deg 82.8$, so $\tan 82.8°$ is $D/568$. Since $\tan 82.8° \approx 7.92$, we conclude that D is approximately 4490 miles. Ridiculous!

Aristotle's argument could be used to dismiss the flat Earth theory, but we can use a little logic to design an experiment that would refute the flat Earth interpretation of Eratosthenes's measurement.

To do this, we will use the logical technique of *reductio ad absurdam*, which is Latin for "reduce to the ridiculous." In this technique, we make successive small changes to the situation and look for a non-

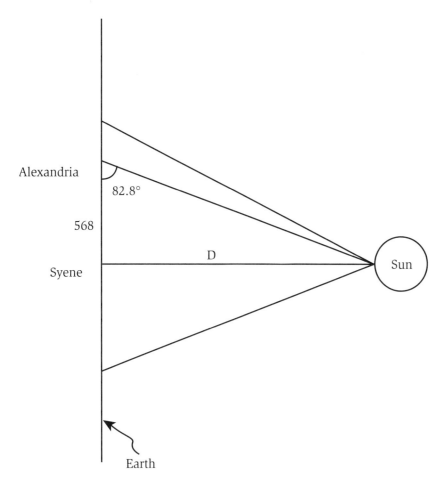

Figure **1-22.** Distance to the Sun from a flat Earth.

sensical result. Continuous small changes could not take us from the realm of the sensical to the realm of the nonsensical, so we must conclude that our original assumption was incorrect.

Assume that the Earth is flat and that Eratosthenes has measured the length of the shadow. Now, ask him to move some distance either toward Syene or away from Syene.[20] This gives us the situation shown in Figure 1-23.

What happens when he is as far from Alexandria as Alexandria is from Syene? This would put him near Istanbul, by the way. Since the distance along the Earth's surface has doubled, and because the actual angle he is measuring is proportional to distance (radian measure), we know that the angle he would measure in Istanbul should be about double the angle he measures in Alexandria.

We can calculate what he would find at Istanbul. Let H_A be the angle he measures at Alexandria (recall that this is about 7.2°), and let H_I be the corresponding angle at Istanbul. We believe that the Earth is spherical, so H_I should be $2H_A$. Let E_A and E_I be the corresponding *elevation* angles; we already saw that E_A is about 82.8°. Looking at the pictures, we see that on a flat Earth

$$\tan E_A = 4490/568 \text{ and}$$
$$\tan E_I = 4490/1136$$
$$= 3.95$$

so $E_I \approx 75.8°$, and therefore $H_I \approx 14.2°$. This is too small; we expect H_I to equal 14.4°.

Charts and Projections

Early navigators actually carried globes to sea. This simplified some of the mathematics involved in navigation,[21] but it is impossible to carry a globe big enough to get much precision. It would take a globe the size of a small building to get the accuracy seen on current aeronautical charts.

Let's work this out. A chart[22] has a scale (which may vary) that is expressed either as a ratio (e.g., 1/500,000, which is sometimes written 1:500,000) or as a relation between different units (this can be

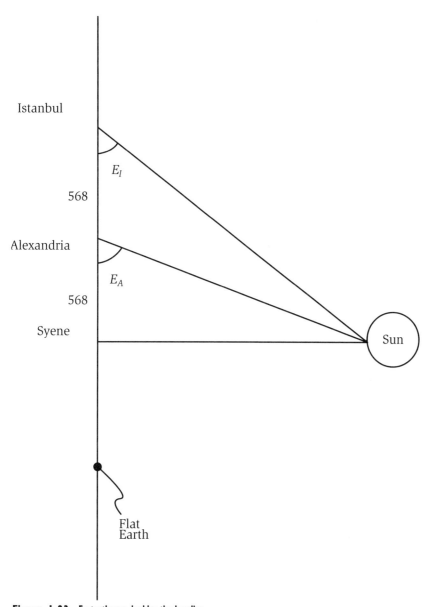

Figure I-23. Eratosthenes doubles the baseline.

something easy, like "1 inch = 1 nautical mile," or something brutal, like "1 inch = 7 nautical miles"). The ratio is a relation between measures using the same unit: 1/500,000 means that 1 inch on the chart corresponds to 500,000″ on the ground. This is 500,000/12 ≈

41,667 feet, or 41,667/6,080 ≈ 6.9 nautical miles. A scale of "1 inch = 7 nautical miles" means that 1 inch on the map is 7 × 6989 × 12 = 510,720 inches on the surface.

Current aeronautical charts use 1:500,000 as a scale, so a globe representing the Earth at this scale should have diameter 1/500,000 times the diameter of the Earth. We saw that the radius of the Earth is 21,600/2π ≈ 3438 nautical miles; this is 3438 × 6080 ≈ 20,903,040 feet; 20,903,040/500000 ≈ 41.8 feet. Thus, to navigate to current standards of accuracy using a globe would require a globe with a diameter of more than 80 feet.

So we use flat charts. A flat chart can't represent the round Earth, so we need to approximate the real Earth. The simplest way to do this is to *project.* In a projection, you take your sphere and a plane that may or may not intersect the sphere. You pick a point C, usually on neither surface. This point is called the center of the projection. To transfer a point P on the sphere to the map, you draw the line $\overline{CP}$ and use the point P' where it hits the plane, as in Figure 1-24.

Notice that all of the lines connecting points on the sphere to points on the map meet at the center of projection C. There is a variation of

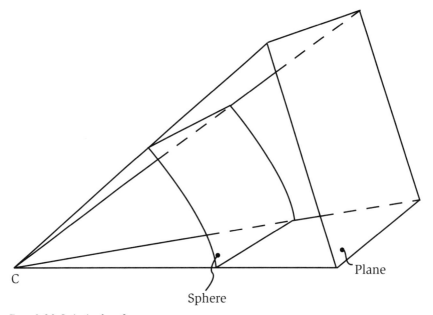

Figure I-24. Projection from C.

projection in which all of these lines are parallel; this is the case with the "forgetful" projection discussed below. In this case, there is no center of projection in the "finite part" of space. Instead, we say that the center of projection is the "point at infinity." The subject of projective geometry explores why we can think of this as an honest point in some space.

When projecting, it can be useful to use a surface other than a plane as the image. For example, both cylinders and cones can be unrolled into flat sheets, as shown in Figure 1-25.

Surprisingly, a *torus* can also be rolled out flat, by cutting it along a horizontal circle and a vertical circle. Keep in mind that the torus is the glaze on the doughnut, but not the doughnut itself. There are not many uses for toroidal projections, as far as I know.

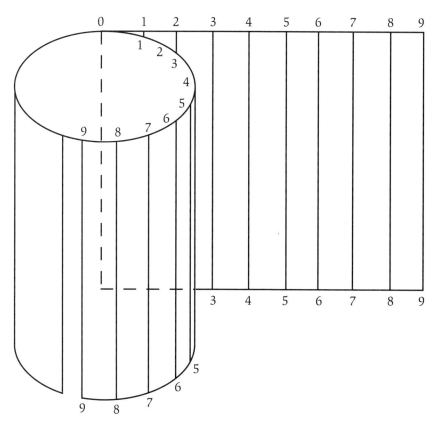

Figure **1-25**. Cylinders and cones are "flat."

Coordinates on a Sphere

We have already discussed how to use x and y or horizontal coordinates on the plane, so we know how the projected points will be represented on a plane. But how do we describe points on a sphere?

One way is to say that the sphere is in 3-space, so we can just use the coordinates in 3-space to describe a point. This is the first calculation that a GPS receiver makes, before converting the result to latitude and longitude. This is a useful perspective, and we will use it a lot. But there is also a lot of waste in it. If we have a point (x, y, z) on a sphere of radius, say, R, whose center is at the origin, then we know that $x^2 + y^2 + z^2 = R^2$, so one of the coordinates is almost redundant. For example, if we know, say, the values of x and y, then we can almost solve for z since

$$z^2 = R^2 - x^2 - y^2 \text{ so}$$
$$z = \pm\sqrt{R^2 - x^2 - y^2}.$$

This would do the trick, except we don't know whether to take the positive or negative value of the square root. Also, since there are points on the sphere with positive values of z and other points with negative values of z, there is no way to make a general rule for picking the sign. If for some reason we knew (or suspected) that z were, say, positive, then we would be able to determine z completely.

There is a projection based on these coordinates that is not used very much in navigation but might help you focus on the ideas. This "forgetful projection" takes the upper (or lower) hemisphere onto the (x, y)-plane by simply forgetting the z coordinate. The image of this projection is a circle whose radius is the radius of the sphere. Notice that for this projection the center of projection is infinity, that is, the center is the "point at infinity." See Figure 1-26.

Polar Coordinates

There are two other common ways of representing points in 3-space, and each has advantages. As usual, we start with a plane and generalize to space.

A point P in a plane has *polar coordinates* that tell you how far the point is from the origin (this is called r) and an angle from the positive x axis (usually called θ, the Greek letter theta). This is illustrated in Figure 1-27. People use polar coordinates all the time in ordinary conversation. "Drive 10 miles east" specifies the direction to go and the distance.

You can see right away that the polar coordinates are related to the ordinary Cartesian coordinates using trigonometric functions. Notice that

$$\cos \theta = x/r \text{ and}$$
$$\sin \theta = y/r.$$

Therefore, solving for x and y,

$$x = r \cos \theta \text{ and}$$
$$y = r \sin \theta.$$

Converting the other way, we know that $x^2 + y^2 = r^2$, and we can see that $\tan \theta = y/x$, so

$$r = \sqrt{x^2 + y^2} \text{ (positive square root)}$$
$$\theta = \arctan(y/x) \text{ except when x} = 0.$$

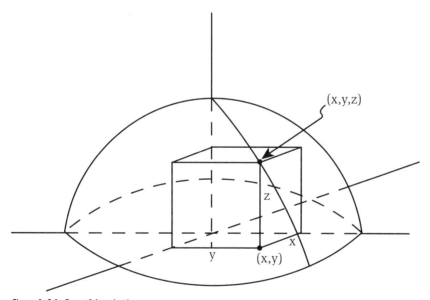

Figure **1-26.** Forgetful projection.

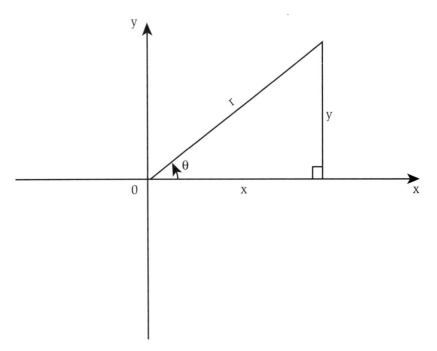

Figure 1-27. Polar coordinates in the plane.

(When $x = 0$, then $\theta = \pi/2$ if y is positive.)

Cylindrical Coordinates

There are two generalizations of polar coordinates that are commonly used in 3-space: cylindrical coordinates and spherical coordinates.

The *cylindrical coordinates* (Figure 1-28) of a point in space consist of the ordinary z coordinate and the polar coordinates of the forgetful projection onto the (x,y)-plane. Thus the conversion in one direction is

$$x = r \cos \theta$$
$$y = r \sin \theta$$

while in the other direction it is

$$r = \sqrt{x^2 + y^2} \text{ (positive square root)}$$
$$\tan \theta = y/x \text{ except when } x = 0.$$

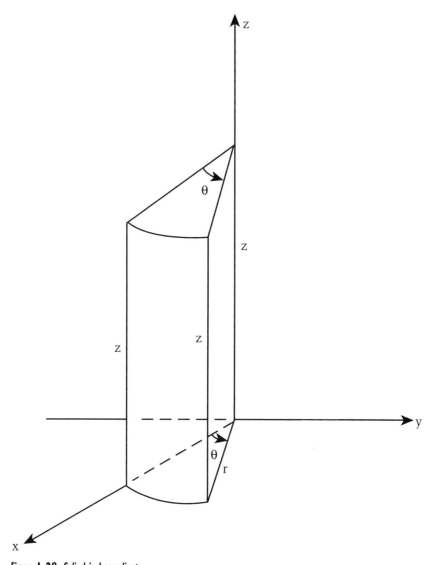

Figure 1-28. Cylindrical coordinates.

In all cases,

$$z = z.$$

The name *cylindrical coordinate* comes from the fact that the solution to the equation $r = r_0$, where r_0 is some constant, is a cylinder.

Spherical Coordinates, Latitude, and Longitude

Spherical coordinates (Figure 1-29) also use a distance, usually called ρ (Greek letter rho). The other two coordinates are angles. One coordinate, θ, is exactly the same θ used in cylindrical coordinates. It measures rotation about the vertical or z axis. The other, called ϕ (Greek letter phi), measures the angle *down* from the North Pole.

How can we convert these to other kinds of coordinates? This is a problem with an interesting twist, because it is easier to solve if we introduce another variable, namely r from cylindrical coordinates. Figure 1-30 is a plane $\theta = \theta_0$. It is clear from the picture that

$$sin\ \phi = r/\rho, \text{ so}$$
$$r = \rho\ sin\ \phi\ \text{ so}$$
$$x = r\ cos\ \theta = \rho\ sin\ \phi\ cos\ \theta\ \text{ and}$$
$$y = r\ sin\ \theta = \rho\ sin\ \phi\ sin\ \theta$$

Finally, $\cos \phi = z/\rho$, so

$$z = \rho \cos \phi.$$

Judging by the name, spherical coordinates are well-suited for studying the geometry of a sphere. On a sphere, ρ is a constant by def-

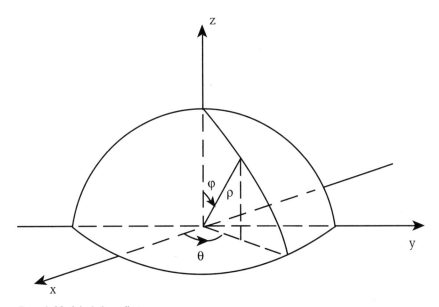

Figure **1-29.** Spherical coordinates.

inition and can therefore be ignored in many calculations. (You only need to know ρ if you want to convert to some other coordinate system.) This means that you can specify a point on the sphere using two numbers, which happen to be angles: θ and ϕ.

This is related to one of the standard methods for specifying a point on the surface of the Earth, namely, using *latitude* and *longitude*.

Longitude is just θ, or at least it is just θ after we have specified a meridian where θ is 0. (Longitude is always between $0°$ and $180°$, and the direction east or west from Greenwich must be stated, while mathematicians allow θ to vary between $0°$ and $360°$; but see below.) The current standard is the meridian through the observatory at Greenwich, England; the meridian is marked on the sidewalk ("pavement" in British parlance) outside. This is called the *prime meridian*.

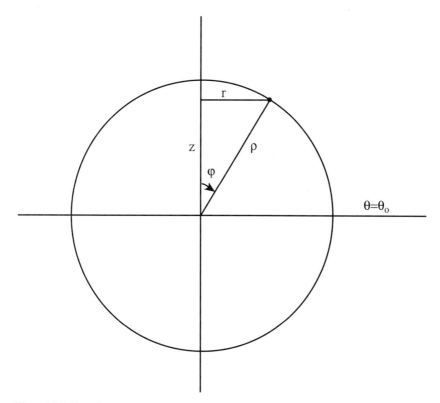

Figure 1-30. Plane slice.

But this is an arbitrary choice, and for many years there was wide-spread disagreement on which meridian should be used as the prime meridian. French navigators used the meridian of Paris; American navigators used Washington, D.C.; and so forth. Some old charts have two longitude scales: I have one marked "Longitude West of Greenwich" across the top and "Longitude West of Washington" across the bottom.

There was no disagreement about the *difference* in longitude between any two points; the difficulty was deciding where 0 should be. To illustrate, Washington, D.C., is about 78° west of Greenwich. Siasconset (pronounced "sconset"), on the eastern tip of Nantucket Island, Massachusetts, is about 8° east of Washington (we are ignoring the latitude in this example), so it is about 70° west of Greenwich (see Figure 1-31).

Miami, Florida, is about 80° west of Greenwich or 2° west of Washington; how far west is it from Siasconset? Comparing the two Greenwich-referenced longitudes, we see that Miami is 80° − 70° = 10° west of Siasconset. Or, comparing the Washington-referenced longitudes (and negating east longitude), the difference is 2° − (−8°) = 10°, again.

If you're bothered about using negative numbers for east longitude, we can approach this from the other direction. Since Siasconset is 8° east of Washington and there are 360° in a whole circle, Siasconset is 352° west of Washington. Comparing the two Washington-referenced longitudes, the difference is 2° − 352°; kind of a mess. But Miami is also 362° west of Washington, since 360° more takes us all the way around the circle. Thus, Miami is 362° − 352° = 10° west of Siasconset. Most astronomers measure latitude this way, as did Captain Vancouver during his exploration of the Pacific Northwest. It's easier to write computer programs this way, too.

The point is that the change of origin adds the same constant to all longitudes, so when you subtract two of them, the constants cancel. Algebraically, $(a + c) − (b + c) = a − b$.

The potential for confusion is very high with two origins. In fact, any time a navigator has to make a choice, there is a pretty good chance that the wrong one will be made. For example, many student

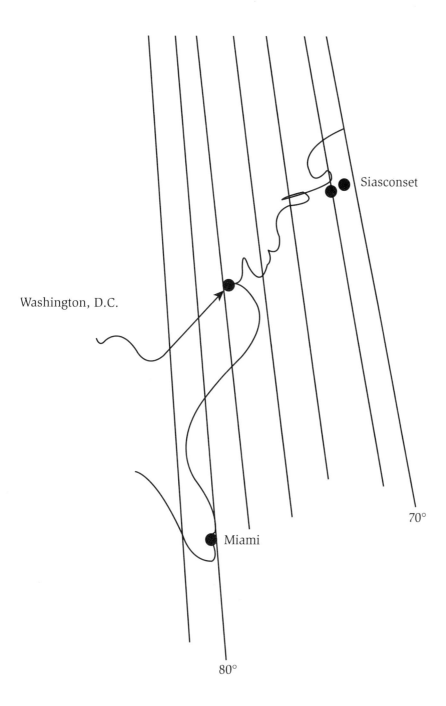

Lambert Conformal Projection

Figure 1-31. Relative longitudes.

pilots buy plotters with four scales given by the permutations of statute and nautical miles for choice of unit and 1:1,000,000 and 1:500,000 (the two most common) for choice of scale. For most flights in small airplanes, the nautical mile with 1:500,000 scale is most useful, but students routinely pick the wrong side of the plotter, especially in flight. Some orange highlighter on the proper side usually helps.

Time Zones

Given this potential for confusion and the international nature of sea navigation, the World Conference on the Meridian was convened in Washington, D.C., in 1884, and a worldwide standard was chosen: Greenwich.

It is interesting to go back and read what contemporaries thought. In (*Science*), the anonymous author wrote that "[i]t was almost a foregone conclusion" that Greenwich be selected because of the great expense and confusion that a new choice of prime meridian (which would require changes to charts and astronomical tables) would cause.

But the conference did more. As was said above, mathematicians measure θ from the prime meridian (which is defined by the x axis) from 0° to 360°, but longitude is measured east or west of the prime meridian. Since 180° West is also 180° East, longitudes vary only between 0° and 180°. This system was one of the conference's recommendations. *Science* opined that the mathematician's method would be less prone to error, but nonetheless agreed with the recommendation.

The reason was another of the conference's topics: the location of the international date line. *Science* wrote "So long as the sun shines . . . the mean solar day . . . must be the great natural unit of time-reckoning. Moreover, for civil purposes the date must change during the hours of sleep. . . . A necessary consequence is, that on some meridian of the globe, . . . the local time must jump one day." The decision was to put the date line at the point where 180° West is also 180° East.[23]

But there was more: the conference recommended that hours should be counted from 0 to 24 over the course of the day; this is in fact done at sea, in the air, and by the military, as well as in many

European countries. The author called the use of A.M. and P.M. in describing time as "clumsy," but noted that these terms would remain common until watchmakers began to manufacture watches with 24-hour dials rather than 12-hour dials. But this would lead to another problem that seems a little quaint to us: repeaters, that is, clocks that strike the hours, would have to strike the bell 23 times at 11 P.M., and this would "seem interminably long."

Ironically, many digital quartz watches give the owner the option of selecting either a 24-hour or 12-hour display, and they "strike" only once each hour (as an owner-programmed option), so it took a century for technology to accommodate the conference's recommendation.

Latitude by Polaris

After all that, it would seem that the other coordinate, latitude, would be easier to define—just use ϕ. This is how mathematicians and physicists work with spheres. Recall that ϕ is measured down from the North Pole and therefore varies between 0 and π radians. In particular, ϕ is always positive.

But ϕ is *not* latitude. Latitude is measured above or below the equator and varies between 90° and −90°. We don't usually use negative values of latitude; instead we say, for instance, that Rio de Janeiro is approximately 23° *south* of the equator.

Why use latitude instead of ϕ? One explanation lies in an odd coincidence for Earth in this epoch: there is a star very close to being directly over the North Pole. This star is called Polaris or the North Star, and beginning in the sixteenth century (if not earlier), navigators used sightings of Polaris to determine approximate latitude. Here's how it works: We assume that all light rays coming from Polaris are parallel, and measure the angle of elevation H of the star above the horizon. This is called its *altitude*, which can be a very confusing term; the same term is used to describe the linear height of an aircraft above the Earth's surface. But the latter is measured in feet while the former is measured in degrees.

The horizon is parallel to the tangent plane through the observer's position P, and we have seen that this plane is perpendicular to the line $\overline{OP}$, where O is the center of the Earth.[24] The perpendicular $\overline{OQ}$ to $\overline{OP}$ through O is parallel to the horizon, so the angle from Polaris through O and Q is, again, H.

Consider the light rays from Polaris to P and from Polaris to the North Pole, N. The latter goes through O and is one side of the angle H. Thus the angle NOP is $90° - H$; but this is $90°$ minus the latitude, so H is the latitude. (Figure 1-32.)

With accurate instruments and a table of the distance between Polaris and the true North Pole, it is possible to measure latitude pretty accurately, at least in the northern hemisphere (Polaris is not visi-

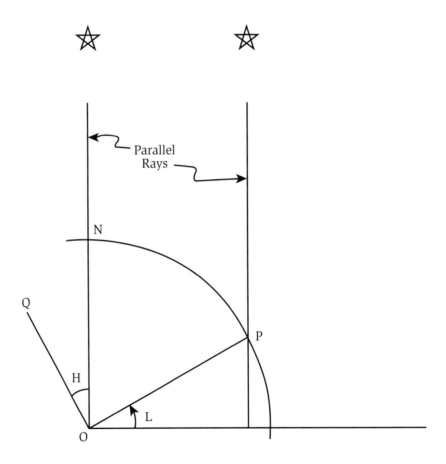

Figure 1-32. Latitude by Polaris.

ble from the southern hemisphere). So for many years, the goal of navigational theory was to find a method to measure longitude, especially while at sea. A big part of this, as we will see later, was accurate time measurement. The basic idea was to compare the local time when the Sun is at your meridian (when its altitude is highest) with the time at the prime meridian; the difference is your longitude. See (Sobel).

It turned out, however, that navigational practice was improved by solving a problem that is more general and therefore sounds much harder than finding longitude. This problem was to determine the set (or locus) of all places on Earth where a certain altitude observation might have been made. For example, suppose you know that the altitude of the Sun in your location was 31° at a certain time. There are plenty of places on the Earth's surface where this could be true; in fact, the set of such places is a circle on the Earth's surface. Knowing the Sun's altitude at a specific time means that you are somewhere on that circle. By intersecting several of these circles, you determine your position. This will be the content of Chapters 3 and 4.

Cylindrical Map Projections

Now that we have tools (namely, coordinates) for working with points on a sphere, we can go back to looking at some of the problems of "mapping" a round sphere onto a flat paper chart. We're only going to solve the problem of mapping one point on the sphere to one point on the chart. To make a useful chart, you need to map many, many points. To make a chart of a coastline, for example, you could pick points that "define" the shore, transfer all of these points to the chart, and draw straight lines in between. If you choose the points wisely, the result will look a lot like the shoreline. In fact, cartographers refer to a list of coordinates of the outline of an area as a "shoreline file." If you want to build a moving map display, you need a shoreline file with the appropriate boundaries (see Figure 1-33).

In the chart, we haven't really "projected" at all. Instead, longitude has become x and latitude has become y. This is an acceptable method for charts of small areas, but at larger scales there are problems

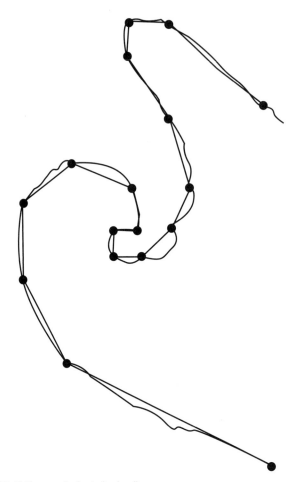

Figure 1-33. Making a crude chart of a shoreline.

because the meridians all meet at the poles. Looking at a globe, you can see that the distance between 10° West and 20° West varies quite a bit. It is fairly far at the equator (about 690 statute miles), but it approaches zero as you get closer to the either of the poles.

Calculating the Projection

Cylindrical projections (Figure 1-34) are among the most useful and easiest to analyze, so we'll start with them. Generally, you take a sphere of radius R with center at O and project onto a cylinder tan-

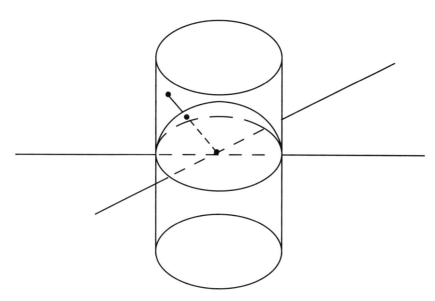

Figure 1-34. Cylindrical projection.

gent to the sphere along the equator. Thus, the cylinder has radius R as well. Its height is irrelevant for our purposes—in fact, we may want to think of its height as being infinite.

As you might imagine, these are easier to look at in terms of cylindrical coordinates. It's very easy to get confused about all the x's and y's floating around, so let's use the subscript s to denote a coordinate in space and the subscript p to denote the coordinates on the plane. Note that the North and South Poles cannot be sent to the cylinder in this way, so we must ignore them.

The coordinate x_p in the plane is just θ (although it might be easier to change the units), and the y_p coordinate in the plane is independent of θ. (We are basically converting radian measure of angles back into distance by doing this.) This means that the only calculation we need to do happens in the plane $\theta = \theta_0$, as in Figure 1-35.

Let P be the point of interest on the sphere, and suppose that P has cylindrical coordinates (r_s, θ_0, z_s). Then using Figure 1-35 and the properties of similar triangles, we see that $z_s/r_s = y_p/R$, so

$$y_p = Rz_s/r_s.$$

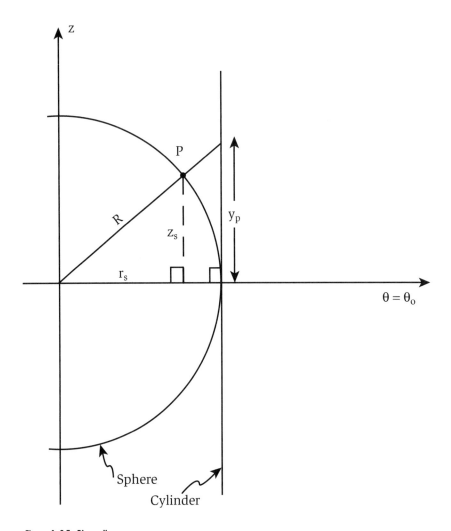

Figure 1-35. Plane slice.

Using latitude and longitude instead of cylindrical coordinates, we see immediately that the coordinate x_p in the plane is just θ, because θ is longitude. The latitude L satisfies

$$\tan L = z_s/r_s,$$

so

$$y_p = R \tan L.$$

Distortion of Cylindrical Projections

A projected map is a distorted map. How distorted? You can think of distortion in several ways. Subjectively, does the map look right? Deciding this means having a picture of what the map *should* look like, which is probably based on another map.

More analytically, there are two things we can measure about a map: distances and angles. It is possible to make maps that preserve angles completely; that is, an angle measured on the map is exactly the same as the corresponding angle measured on the Earth. Such maps are called *conformal*. This is obviously an important trait for a chart used for measuring courses and bearings. A Mercator chart is conformal, as is a Lambert chart (which is the primary chart used in air navigation).[25]

The cylindrical projection we have been discussing is not conformal, although we won't be able to prove this fact until we have developed more tools (especially tools for working with great circles). How does the cylindrical projection distort lengths? We can measure this by considering the ratio of projected distance to actual distance. That is, let P_1 and P_2 be two points on the sphere. For convenience, we'll assume that they are on the same meridian (we'll need the great-circle tools mentioned above in order to measure arbitrary distances on the sphere, so we'll concentrate on this special case).

Let L_i denote the latitude of P_i, and suppose that $L_2 > L_1$.[26] As usual, latitude is measured in degrees. Since latitude is an angle, we can use radian measure and convert the difference in latitude directly to distance. See Figure 1-36. The radian angle is $\pi(L_2 - L_1)/180$, and the actual distance D_s is thus

$$D_s = \frac{\pi R(L_2 - L_1)}{180}$$

where R is the radius.

We know that the projected point at latitude L_i has y- coordinate $R\tan L_i$ on the map, so the map distance D_p is given by

$$D_p = R(\tan L_2 - \tan L_1).$$

Thus, the distortion factor is

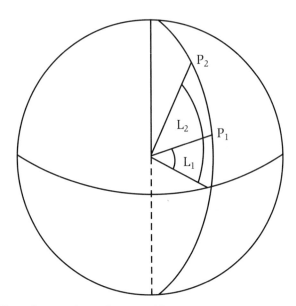

Figure 1-36. Distance between points on the same meridian.

$$\frac{D_p}{D_s} = \frac{180(\tan L_2 - \tan L_1)}{\pi(L_2 - L_1)}.$$

This expression is hard to interpret; its graph is a surface above a plane with axes labeled L_1 and L_2. Table 1-2 shows values as the Ls vary from 0° to 75° in increments of 15°. Notice that the "southwest" portion of the table is empty, since $L_2 > L_1$. In the table, read L_2 across the top and L_1 down the side.

Table 1-2. Distortion in a Cylindrical Projection

–	15°	30°	45°	60°	75°
0°	1.03	1.09	1.26	1.66	2.86
15°	–	1.20	1.38	1.89	3.32
30°	–	–	1.60	2.18	4.01
45°	–	–	–	2.75	5.21
60°	–	–	–	–	7.62

When L_1 is near zero (when one of the points is near the equator), the distortion factor is fairly small; the mapped distance from the equator to 15° North is approximately 1.03 times the actual distance. But the same distance on the globe is distorted by a factor of almost 8 when one is near 60° North and the other is 75° North.

Another aspect of a cylindrical projection is worth mentioning. On the globe, meridians get closer as one approaches the pole; this is not the case on the chart. Thus, there is horizontal distortion as well. The distortion factor is approximately the cosine of the latitude. Thus, if two meridians are 1 unit apart at the equator, the cylindrical projection shows them 1 unit apart until the top off the map, but they are actually $\cos 30° = 0.8660$ units apart at latitude 30°, and $\cos 60° = 0.5$ units apart at latitude 60°.

A *transverse projection* is a cylindrical projection where the circle of tangency is closer to the points of interest, rather than at the equator. These are widely used.

Mercator's Chart

A projected map is a distorted map. This sounds negative. But perhaps the distortion can be put to good use? This is the case with Mercator's projection, which is a cylindrical projection with a scale factor applied to the vertical component. Its full analysis requires calculus, but the net result is that a straight line on a Mercator chart is a *rhumb line* on the globe; that is, it crosses each meridian at the same angle. Since the meridians converge at the pole, this means that the rhumb line looks like a spiral whose center is the pole. Since the horizontal distortion is the cosine of the latitude, the north coordinate of each little segment must be divided by the cosine of the latitude to preserve the rhumb line property. Using calculus, one can then determine how far up the chart each latitude should be pushed.[27] An approximate version is presented in Chapter 3.

Mercator charts aren't used much in aviation anymore, although Lindbergh used one for navigation during his 1927 flight.[28] The Lambert conformal chart, which is a variant of the conic projections discussed in the next section, is much more convenient. The various transverse Mercator charts are widely used in land management.

Conic Projections

We saw from the distortion table that a cylindrical projection gets more distorted as you move away from the equator. One way to try to reduce this distortion is to project onto a surface that stays closer to the sphere as you move north. For example, a cone whose axis is aligned with the sphere's axis stays closer than a cylinder does, as shown in Figure 1-37.

Such a cone is tangent to the sphere along a circle parallel to the equator. We will choose the center of the sphere O as the center of projection, which means that meridians are sent to lines through the vertex. We will then slice the cone open along one of these lines to make a flat map.

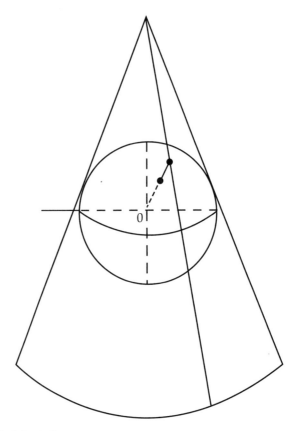

Figure **1-37**. Conical projection.

The problem here is that θ on the sphere no longer becomes x_p on the map: the image of a circle of constant latitude from the sphere is a circular arc on the map, *not* a line, as was the case with a cylindrical projection.

To figure out how to handle this, we need to consider the geometry of the cone a little more closely. The cone can be specified by two dimensions: its base radius r and its height h. Since we are only considering so-called right cones, whose axis is perpendicular to the base, when we slice the cone open along a line through the vertex we get a segment of a circle whose radius is $\sqrt{r^2 + h^2}$ (Figure 1-38).

When the cone is sliced open, the base becomes a circular arc whose arc length is $2\pi r$. Thus we can measure the angle V at the vertex. In radians, the measure is

$$V = \frac{2\pi r}{\sqrt{r^2 + h^2}},$$

which is $V = 360r/\sqrt{r^2 + h^2}$ degrees.

Now (spherical or cylindrical) θ on the sphere varies from $0°$ to $360°$, and each meridian $\theta = \theta_0$ corresponds to a line through the vertex, so as θ varies these lines move across the sliced-open cone like a windshield wiper pivoting on the vertex. This means that we cannot determine x_p on the map from θ alone, as we could with the cylindrical projection.

Coordinates on the Cone

Let A be the latitude of the circle of tangency between the cone and the sphere. Each meridian on the sphere becomes a line on the cone, and we need to specify coordinates on these lines. There are an infinite number of choices for the origin and positive direction of these coordinate systems, but two stand out: let the origin be the vertex (in which case we can take all coordinates to be positive), or let the origin be the point of tangency and let the positive direction be toward the vertex (Figure 1-39).

We choose the latter. Now consider a point P on the sphere at latitude L. The circle of tangency meets the meridian of P at some point

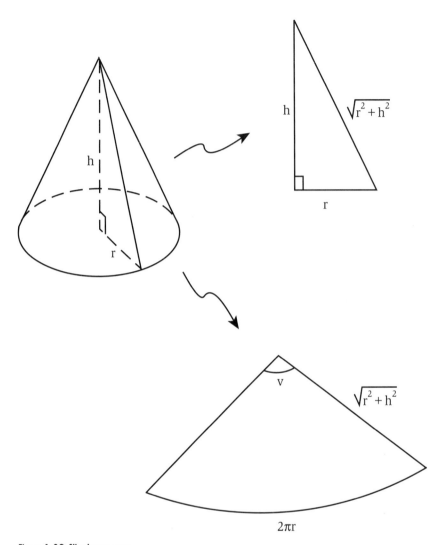

Figure 1-38 Sliced-open cone.

T (which is the origin on the image line), and the angle TOP is $L - A$.

The image point is at some distance D from T; since the image line is tangent to the sphere, it is perpendicular to the line OT, so we can use trigonometry to see that

$$D = R \tan(L - A)$$

where, as usual, R is the radius of the sphere.

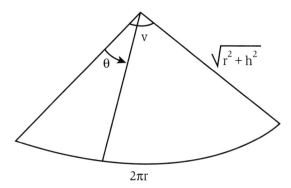

Figure 1-39. "Natural" coordinates on the cone.

We won't work out the exact formulas for plotting the image of a point on the plane, but will leave it with this: if we slice the cone open and lay it out on the table, we can use the windshield-wiper effect to determine the angle (at the vertex) along which the image lies; and, if we measure along the windshield wiper from the circle of tangency, we can determine where on the line (and thus on the paper) the point's image lies. This is like a modified system of polar coordinates in which we use distance from a fixed circle (positive or negative) rather than distance from the origin.

Distortion Factors for Conical Projections

We can, however, make some estimates of the distortion of a conical projection along a meridian. If we have points P_1 and P_2 at latitudes L_1 and L_2, the actual distance between them is still

$$D_a = \frac{\pi R(L_2 - L_1)}{180}$$

but now the projected distance is $D_m = R(\tan(L_2 - A) - \tan(L_1 - A))$. We can almost use the same table of distortions as for the cylindrical projection, but now we read the lines corresponding to $L_i - A$. There is a potential problem if one or both of the L_i is closer to the equator than A; then $L_i - A$ is negative, so we have to negate the value of the tangent.

Here are some examples to make this clearer. First, suppose that $A = 10°$, $L_1 = 25°$, and $L_2 = 40°$. Then $L_2 - A$ is 30° and $L_1 - A$ is

$15°$, so the distortion factor, taken from Table 1-2, is 1.20. However, if $A = 40°$, $L_2 = 55°$, and $L_1 = 25°$, then $L_2 - A$ is still $15°$, but now $L_1 - A$ is negative, and we can no longer use the table. Going back to the formula, the distortion is

$$\frac{D_p}{D_s} = \frac{180(\tan(L_2 - A) - \tan(L_1 - A))}{\pi(L_2 - L_1)}$$
$$= \frac{180(\tan 15° - \tan(-15°))}{\pi(30)}$$
$$= \frac{180(0.268 - (-0.268))}{94.2}$$
$$\approx 1.02.$$

In this case, the distortion is comparable to distortions near the equator for a cylindrical projection, even though L_2 is more than halfway from the equator to the pole. In general, if L_1 and L_2 are both near A, the distortion is fairly small.

Charts in Aviation

Current aeronautical charts use the Lambert conformal projection, which is a variation of the conical projection in the same way that a Mercator chart is a variation of a cylindrical projection. Its analysis requires calculus. In the northern United States and southern Canada, the cones have an "effective tangent" of 44° North. A swath 15° on either side of 44° North includes both San Antonio, Texas, and Juneau, Alaska, as well as Canadian territory as far north as Lake Athabasca. (See Chapter 5 for more specifics on charts.)

As the name implies, the Lambert *conformal* chart preserves angles, so courses can be measured directly on the chart. But since the chart is conical, the meridians converge on a vertex, so a long east-west course visibly crosses successive meridians at different angles. In fact, a straight line on a Lambert chart is almost identical to a great-circle course. Thus, to follow a course plotted on a Lambert chart exactly, a navigator would need to have a constantly changing compass heading. In fact, navigators use rhumb-line approximations to make life easier, with no significant sacrifice of accuracy. See the section on great circle routes in Chapter 3.

Three Kinds of Latitude

So far, we have been assuming that the Earth is a perfect sphere, which is pretty reasonable for some things, but it is not correct. The Earth is a spheroid, and navigators, surveyors, and geographers have put a lot of effort into determining its exact shape.

Sir Isaac Newton, the inventor of calculus and the discoverer of the law of universal gravitation, calculated that the Earth should actually be an oblate spheroid; that is, its diameter at the equator should be a little more than the diameter at the pole. Others, notably Cassini, thought that the Earth should be a prolate spheroid. The difference is illustrated in Figure 1-40.

How could anyone disagree with Newton's calculations? There was no mistake in his calculations; nor was there any mistake in Cassini's. The problem is one we have already bumped into. Even though Newton's calculations were correct, they were based on

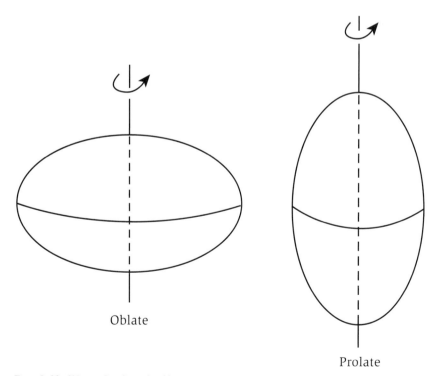

Oblate

Prolate

Figure **1-40. Oblate and prolate spheroids.**

assumptions about the makeup of the Earth and as such needed experimental verification. Other assumptions would lead to different shapes from correct calculations.

The French Academy of Sciences, under Louis XV, used measurements by two expeditions to determine the actual shape of the spheroid.[29] One expedition went north to Lapland, and the other went to the Andes, near the equator. The Lapland party had an easy time of it, since the frozen rivers made it easy to measure distances. This was not the case in the Andes, and it took quite some time for the results (which, combined with the Lapland measurement, confirmed Newton's view) to reach Paris.

By the way, these were a combination of astronomical and ground-based measurements. The interpretation of the ground-based measurements is based on more trigonometry, which will be developed in Chapter 2.

Measuring Latitude

We need to investigate the actual shape of the Earth, but once we have decided that the Earth is not a perfect sphere, we have to put more thought into the problem of measuring latitude. Ironically, longitude, which was the harder measurement to make, is not as sensitive to the shape of the Earth. This is partly because our standard for measuring longitude—time—tacitly refers to a single point on the rotating surface. But there are several ways of defining latitude.

Let's go back to Eratosthenes' measurement of the Earth's circumference. In that discussion, we mentioned several ways of determining "straight up": the direction of gravity (a plumb line), the direction of a line from the Earth's center, and the line perpendicular to the plane tangent to the Earth at the point in question.

On a perfect sphere, these measurements are all the same. Remember that the radius points to the center and is perpendicular to the tangent plane. Thus, the last two methods mentioned give the same result on a perfect sphere. A plumb line is oriented with respect to gravity, and it is a fact that the strength of gravity varies over the surface of the Earth. Part of this is due to Newton's Universal Law of Gravitation, which states that gravitational force is proportional to the inverse square of the distance between the objects in question. Thus, on top of a mountain, you are farther from the center of the Earth, and hence the pull of gravity is a little weaker than it is at the base of the Marianas Trench. The rotation of the Earth induces centrifugal effects that are measurable; they are strongest near the equator. In fact, they vary with the cosine of the latitude.

But even after taking these known effects into account, there is a problem because, as we mentioned earlier, the *density* of the Earth varies from place to place. For example, the Midcontinent Positive Gravity Anomaly in North America is associated with iron ore deposits near the surface.[30] Because of this, gravity is stronger in Minneapolis than it is in St. Louis.[31] Thus, a plumb line in St. Louis is deflected, almost imperceptibly, toward the anomaly.

Here, then, are three definitions of the latitude of a point P on the surface of a planet whose center is O:

1. *Geocentric latitude* (Figure 1-41) is the angle formed by the equator and the ray $\overline{OP}$. It is denoted L_G.

2. *Astronomical latitude* (Figure 1-42), denoted L_A, is the angle formed by the equator and a plumb line, that is, gravity.

3. *Geodetic latitude* (Figure 1-43) is the angle formed by the equator and the normal to the tangent plane at P (this ignores local terrain effects). It is denoted L_N.

The discussion above implies that on a perfect sphere, $L_G = L_N$. At any point on an oblate spheroid, $|L_N| \geq |L_G|$, with equality occurring only at the poles and along the equator. On a prolate spheroid, the inequality is reversed: $|L_G| \geq |L_N|$.

For the purposes of navigation, L_A appears to be useless, especially if you are going to use the stars and planets to help determine your position. This is because gravitational anomalies make the observation dependent on variables other than the stars. Worse, since gravitational anomalies move, the value of L_A changes over time, even if you stay firmly planted at the same spot on the Earth (this is assuming movement below the crust, not plate tectonics). It doesn't seem to be a very reliable measure.

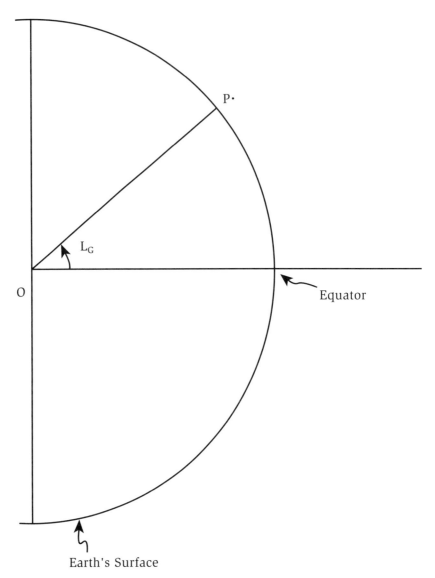

Figure 1-41. Geocentric latitude, or L_G.

However, L_A is a measurement we can make, so it is needed for some applications. Usually, a sextant (the tool for measuring the altitude of a star or planet) needs a horizontal reference, and the typical situation is that the horizon is used for the horizontal reference. But

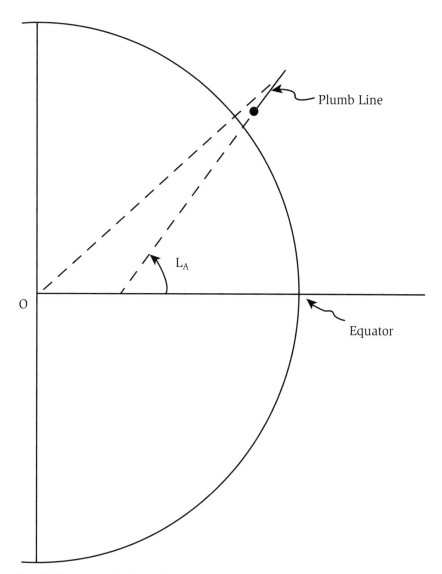

Figure 1-42. Astronomical latitude, or L_A.

what happens if you can't see the horizon? This could easily be the case in an airplane over a cloud deck (and many pilots have had disorienting experiences with sloping cloud decks, so you can't use them for reference). For these situations, navigators use a bubble sextant, which has a small bubble in a liquid, like a carpenter's level. The bubble moves to the "top" of the liquid,—that is, it moves "straight up"

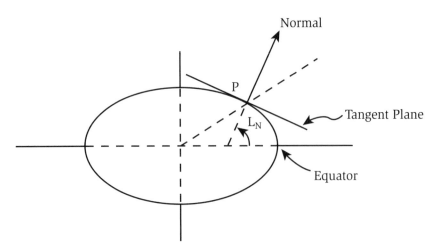

Figure I-43. Geodetic latitude, or L_N.

in the liquid—and it is therefore measuring with reference to L_A. Luckily, the effects of the gravitational anomalies are small—their detection requires very delicate and expensive instruments—so there is no measurable effect on the bubble. On the other hand, there are measurable effects on the bubble from acceleration and the Coriolis effect,[32] and navigators using these tools must consult correction tables in order to remove these errors from the observations.

Ellipses and Their Tangents

The spheroids of interest can be constructed by taking an ellipse and looking at the volume swept out when the ellipse is rotated around one of its axes. The standard equation for an ellipse is

$$\frac{x^2}{a^2} + \frac{y^2}{b^2} = 1$$

where x and y are the usual coordinates and a and b are constant. *Standard* means that any ellipse can be given an equation in this form by proper choice of coordinates. We will always rotate these ellipses around the y axis. If $a > b$, we get an oblate spheroid, and if $a < b$ we get a prolate spheroid (Figure 1-44). The longer of the two defines the *semimajor axis*, and the ratio of their difference to the larger is called the *flattening*. If it is close to zero, then the ellipse is almost a sphere.

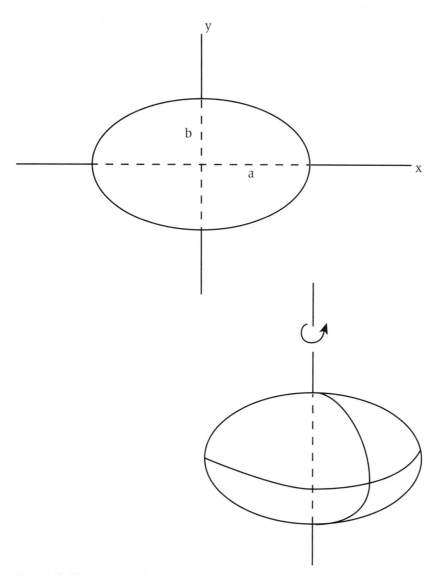

Figure 1-44. Ellipses and spheroids.

For the WGS-84 model of the Earth, the semimajor axis is $6,378,137.0$ meters, the semiminor axis is $6,356,752.3142$ meters, and the flattening is $1/298.257223563$.

The ellipse can be *parametrized* with parameter t by setting $x = a \cos t$ and $y = b \sin t$. You can check right away that $(a \cos t, b \sin t)$ is a point satisfying the ellipse's equation.

A point P on the spheroid can be considered to be a point on the generating ellipse by slicing the spheroid with the plane corresponding to the point's longitude (or θ in the point's spherical or cylindrical coordinates). We can find L_G, the geocentric latitude, immediately: the ray defining the latitude goes through the point $(a\cos t, b\sin t)$, so it has slope[33]

$$\frac{b\sin t}{a\cos t}$$

and this is the tangent of the latitude, that is

$$\tan(L_G) = \frac{b\sin t}{a\cos t}.$$

So far we haven't used any calculus. To find L_N, we need to know slope of the tangent line to the ellipse at P, and using calculus one sees that the normal has slope $a\sin t/b\cos t$; therefore

$$\tan(L_N) = \frac{a\sin t}{b\cos t}.$$

If $a > b$, then it is obvious that $L_N > L_G$, as claimed.

The Shape of the Earth

Now that we know more about latitude, we can examine two mathematical models of the Earth's actual shape. We will examine data from the models and see what they tell us: oblate or prolate?

Consider a really oblate spheroid; for example, suppose that $a = 100$ and $b = 1$. Consider how far you have to travel along the ellipse in order to make a 1° change in L_G, the geocentric latitude (the one determined by drawing a line from the center of the planet to the point in question). Near the pole, this distance isn't very far; basically, you are subtending one degree of arc close to the center, so the "effective radius" is small (and remember that radian measure of the angle is the ratio of arc to radius). Near the equator, however, you are much farther from the center, so the same angle subtends a longer arc. Thus, on an oblate spheroid, the arc defined by one degree of L_G decreases as you move from the equator to the pole. See Figure 1-45.

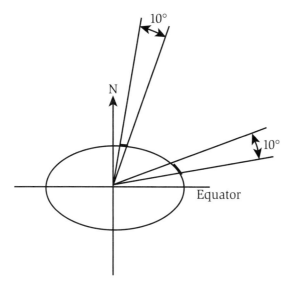

Figure 1-45. Varying L_G on an oblate spheroid.

Using a similar argument, you can see that on a prolate spheroid, one degree of L_G cuts out larger and larger arcs as you move from the equator to the pole.

In practice, L_G is very difficult to measure. Can you point to the center of the Earth from where you are sitting now? The best you could do would be to use a plumb line, but when you do this you are measuring L_A, not L_G, and therefore you are introducing an error whose magnitude is small but probably unknown.

So, how does the arc defined by one degree of L_N vary as you move from the equator to the pole? On the really oblate spheroid (Figure 1-46), you see that near the equator a very small step toward the pole leads to a large change in the tangent plane, and this means a large change in the normal direction used to define L_N.

Near the pole, however, the ellipse appears nearly straight (Figure 1-47), so you need to move a long way along it to effect a 1° change in the tangent (and, therefore, normal) direction.

Thus, on an oblate spheroid, the arc defined by one degree of L_N *increases* as you move from the equator to the pole.

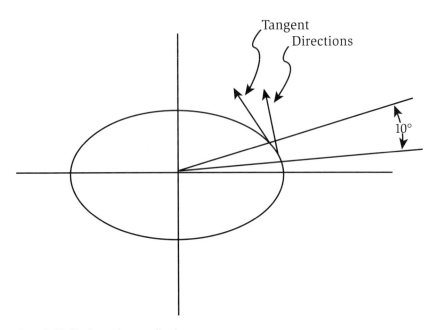

Figure 1-46. Big change of tangent direction.

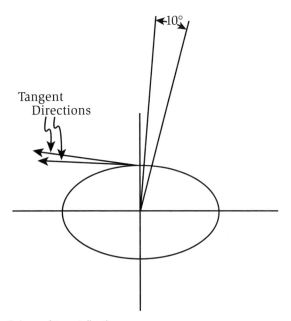

Figure 1-47. Small change of tangent direction.

The Clarke Spheroid

Some geographers may think that the Clarke spheroid of 1866, shown in Table 1-3 and discussed in (Bowditch), is too old-fashioned a model, but it serves our purposes well. It was the standard model of the geoid until the introduction of WGS-84, the World Geodetic Standard of 1984, which will be presented next. Table 1-3 uses nautical miles. The Clarke spheroid notes the difference between the spherical model and the actual shape; it is, we emphasize, a model.

Without the model, a position obtained as (x, y, z) coordinates in space, as would be derived from Global Positioning System satellite observations, cannot be used to determine the elevation. The model tells how far the surface, that is, *sea level*, is from the center of the Earth, while the (x, y, z) coordinates determine how far the observer is from the center of the Earth. The difference between these two is our best estimate of the observer's elevation.

The table contains the length, in nautical miles, subtended by one degree of geodetic latitude (L_N). The model describes the Earth as an oblate spheroid. Notice that a degree of arc is 0.612 NM longer at the pole than it is it the equator. This is a difference of about 1 percent; put another way, 0.612 NM is about 3700 feet.

WGS-84 Spheroid

By contrast, Table 1-4 is WGS-84, the current model, from (Hunerfelt). The maximum difference between these entries and the corresponding entries for the Clarke spheroid is 36 feet, or about 12 meters.

Conclusion

Newton contemplated the shape of the Earth sitting in his study in Cambridge, and he was able to deduce the fact that the Earth is an oblate spheroid. The French Academy spent a lot of money to make the measurements that confirmed Newton's theory. Then Clarke and others contemplated the measured data and determined a curve that fit them well. These models are products of both action and contemplation.

Table 1-3 The Clarke Spheroid

Latitude	Arc
0°	59.701
5°	59.706
10°	59.720
15°	59.742
20°	59.773
25°	59.810
30°	59.853
35°	59.902
40°	59.953
45°	60.006
50°	60.059
55°	60.111
60°	60.159
65°	60.203
70°	60.241
75°	60.272
80°	60.294
85°	60.308
90°	60.313

Table 1-4 WGS-84

Latitude	Arc
0°	59.705
5°	59.709
10°	59.721
15°	59.748
20°	59.772
25°	59.808
30°	59.860
35°	59.908
40°	59.949
45°	60.001
50°	60.055
55°	60.105
60°	60.153
65°	60.197
70°	60.236
75°	60.266
80°	60.289
85°	60.304
90°	60.310

NOTES

1. There are also other operational and proposed satellite systems, but we will generally refer to all of them as *GPS* with no slight of the others intended.
2. Nobody ever mentions the evident roundness of the Moon.
3. Usually, this means the meridian through Greenwich, England; we'll say more about it later.
4. One of the angles is a right angle; that is, its measure is 90°.
5. There is an unwritten and oft-broken convention that variable names from the beginning of the alphabet are constants, and those from the end are variables. So what are h and k?
6. Sometimes the word *simplify* is misused, but in this case the result really is simpler.
7. This method of working with tangents goes back to Descartes.
8. The exception is that a plane is tangent to itself.
9. Geologists call this a *face pole*.
10. This notation is a little bit different from the notation we used above in discussing Pythagoras's theorem.
11. The so-called trivial ratios, like opposite / opposite, are all equal to 1.
12. The usual practice is to land into the wind, since this reduces the speed over the ground at touchdown.
13. A *knot* is 1 nautical mile per hour. A nautical mile, which is about 6080 feet, is a convenient measure for navigation purposes. Knot is itself a measure of speed; knots per hour is a measure of acceleration.
14. Runway numbers are usually determined by rounding the runway's magnetic heading divided by 10.
15. In fact, sin 2A is 2 sin A cos A.
16. Since the sine of the distance is a distance along a great circle, it is the radian measure of an angle.
17. Of course, the problem of finding relative longitude was particularly vexing until fairly recently, too.
18. See the articles on Cook and on transits of Venus in (*Encyclopaedia Britannica*).
19. This is a simplification because it ignores the fact that observers at different locations see Venus cross the Sun at different solar latitudes, so the time the planet crosses the solar disk is later for some observers than for others.
20. Ignore for the moment the fact that moving away from Syene takes him into the Mediterranean Sea, which makes this argument all wet.
21. Every book I know on the history of navigation asserts that most navi-

gators preferred to do little or no math, but Cook and Vancouver, among others, were quite adept at mathematics. See (Raban).

22. A *chart* is a map used for navigation. You can use a chart as a map, but you may not be able to use a map as a chart, although Lindbergh used railroad maps to navigate from San Diego to New York before his 1927 Atlantic flight.

23. Or, at least, approximately there. Variations were allowed so that the islands of Fiji, which are separated by this line, could all be on the same day.

24. The assumption that the horizon is parallel to the tangent plane is clearly false if the viewer is elevated above the surface. This is why there are tables of sextant corrections for *dip* or *height of eye*.

25. There will be more information on these charts in Chapter 5.

26. We can just switch the names of the points if it happens that L_1 is larger. After all, we picked the names.

27. For those in the know, the integral of the secant.

28. As we will see in Chapter 5, he used a different kind of chart for planning.

29. (Wilford), Chapter 7.

30. See (Chernicoff).

31. One has to speculate whether this was part of the reason that the Lakers, playing a highly gravity-influenced sport, moved from Minneapolis to Los Angeles.

32. In the northern hemisphere, a moving object appears to turn to the right, which can cause the bathtub drain to form a counterclockwise vortex as well as the characteristic counterclockwise flow around regions of low atmospheric pressure. Plumbing design makes the latter a more reliable indicator of the Coriolis effect than the former.

33. *Slope* is, as usual, rise over run; we will discuss this in Chapter 2.

CHAPTER 2

Vectors and Spheres

So far, all of the mathematics we have looked at has been direct-ly motivated by problems in either navigation or geodesy. But there have been a few times we have had to say "more on this later" because a lot of technical tools were needed to discuss the matter fully. Now we're going to develop those tools.

Thus, this is the most purely mathematical part of the book. Everything done in this chapter will be used to solve a problem in navigation. All of the material is standard, but the examples come from navigation, and the emphasis is on vectors in space. The biggest example is a formula for determining great-circle distances.

The tools we need are summed up in one word: *vectors*. The sim-plest definition of a vector is "an entity with a direction and a mag-nitude." For example, "Bangkok is approximately 1100 kilometers north of Kuala Lumpur" is a vector statement; vectors like this are called *displacement vectors*. Here's another example: "I flew a com-pass heading of 230° at an airspeed of 148 mph."[1] Here the direc-tion is specified as a particular angle, and the speed is the magni-tude.

Two vectors are equal if and only if they have the same direction and the same magnitude. This makes sense. Thunder Bay is about 1100 kilometers north of St. Louis; thus, the Thunder Bay–St. Louis displacement vector equals the Bangkok–Kuala Lumpur displace-ment vector.[2]

There is one exceptional vector whose length is zero and whose direction is undefined. This is called the *zero vector*.

If this was all there was about the subject of vectors, they would not be worth the bother. But there's more, and it's another situation where mathematical abstraction leads to useful and interesting tools.

Basic knowledge about vectors dates back at least to Euclid, but nowadays people learn about vectors in either physics or advanced calculus. This is a shame, since such a useful tool should be more widely understood.

Vector Algebra

The word *algebra* usually brings to mind long lists of quadratic polynomials to factor and other arithmetical operations with unknown quantities. In fact, that's a good definition. Algebra is the study of general arithmetical properties of numbers. The "unknown" is often a "don't care": I don't care which real number x denotes, because it is always true that if you square x and subtract 4, you get the same result as if you first subtract 2 from x and multiply that by the result of adding 2 to x. (This is a very long-winded way of saying "for all real numbers x, $x^2 - 4 = (x - 2)(x + 2)$.")

The thing about algebra is that we are so conditioned to think of the symbols as having a fixed meaning that we lose sight of the fact that what we are studying is relationships, not individual numbers. This is an important distinction to the navigator, one of whose jobs is to figure out the relationship between two points on the Earth's surface.

Consider addition, " + ". It has some properties, and these are so important that they seem self-evident. For example, whatever a and b represent, $a + b = b + a$. Similarly, $a + (b + c) = (a + b) + c$, whatever a, b, and c might represent. There is a special number called zero (*0*); adding 0 to any number gives you the same number back. And, given a number a, there is another number b such that $a + b = 0$; we usually call b something related to a, namely $-a$.

What does this have to do with vectors? First, let's represent vectors in the obvious way by drawing them as arrows; the length of the arrow is the magnitude of the vector,[3] and the direction the arrow points is the vector's direction. Remember, two arrows that point the same way and have the same length represent[4] the same vector.

It is convenient to have a visual way of telling you that a certain quantity is a vector quantity. Logically, all I have to do is write "Let A be a vector," and leave it up to you to remember. But it's easier for everyone if I instead write "Let **A** be a vector," and let the boldface act as a reminder. So we'll use this convention throughout.

Define a vector operation called the *resultant* of two vectors, as illustrated in Figure 2-1. Let **a** and **b** be vectors, drawn as arrows. Move **b** around on the page until the tail of its arrow is exactly at the head of the arrow representing **a**.

This defines a new vector, called the *resultant* of **a** and **b**, going from the tail of **a** to the head of **b**. The resultant is denoted using the familiar symbol of horizontal and vertical line segments joined at their midpoints: $+$. The reason we use the same symbol that we use for everyday addition is that the resultant has the same properties: if $\mathbf{a}_1$, $\mathbf{a}_2$, and $\mathbf{a}_3$ are any three vectors, then $\mathbf{a}_1 + \mathbf{a}_2 = \mathbf{a}_2 + \mathbf{a}_1$, $\mathbf{a}_1 + (\mathbf{a}_2 + \mathbf{a}_3) = (\mathbf{a}_1 + \mathbf{a}_2) + \mathbf{a}_3$; there is a vector **0**, namely the zero vector, which gives you back the vector you started with when added; and each vector has a negative, which is produced by keeping the magnitude and reversing the direction.

Adding Vectors

Even with all of these similar properties, vectors are not numbers. One of the big differences is in multiplication. We're going to define three different multiplications involving vectors. All of them share several

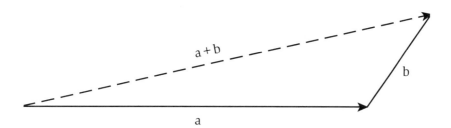

Figure 2-1. Vector addition.

properties with ordinary multiplication of numbers; this legitimizes the use of the word *multiplication* to describe the operation. But none of these multiplications will share *all* of the properties of ordinary multiplication.

There's one thing we have to be clear on: If we have three kinds of multiplication, we always have to be careful to use the one we mean to use. Actually, there is little chance of confusion as long as you pay attention to what each symbol represents. Table 2-1 shows the properties of these three products. We will develop each one after we know how to represent vectors in coordinates.

Scalar Multiplication

The first of these products—the *scalar* product—changes the length of the vector, but not the direction. Most people write the length of a vector **v** using absolute value signs: the length of **v** is denoted $|\mathbf{v}|$. The scalar product takes a number r and a vector **v**, and produces a vector with the same direction as **v** but whose length is $r|\mathbf{v}|$. For example, $2\mathbf{v}$ has twice the length of **v**, $\pi\mathbf{v}$ has length π times $|\mathbf{v}|$, and $-3\mathbf{v}$ is 3 times as long as **v**, is parallel to **v**, but points in the opposite direction. This is illustrated in Figure 2-2.

Notice that there is no particular symbol for scalar product. Just write the number next to the vector; this is a lot like the way we write ordinary multiplication of numbers represented by single letter symbols. But when you multiply two numbers, you get a number; the scalar product takes a number and a vector and produces a vector.

Of course, if **v** is a vector then $0\mathbf{v}$ is the zero vector, **0**, because its length is zero.

Table 2-1. Trigonometric Functions

NAME	FACTOR 1	FACTOR 2	RESULT
Scalar	Number	Vector	Vector
Dot	Vector	Vector	Number
Cross	Vector	Vector	Vector

The name *scalar* comes from the fact that all this operation does is effect a change of scale. Some people, when working with vectors, refer to numbers as "scalars."

Here are some properties of scalar product. In these formulas, **v** and **w** are any vectors and r and s are any numbers.

$$(r + s)\mathbf{v} = r\mathbf{v} + s\mathbf{v}$$
$$r(\mathbf{v} + \mathbf{w}) = r\mathbf{v} + r\mathbf{w}$$
$$(rs)\mathbf{v} = r(s\mathbf{v}).$$

If you ignore the boldface letters, these are properties of ordinary multiplication.

Vectors in Coordinates

Arrows are too hard to work with analytically, so we are going to introduce coordinates and derive some formulas for vector operations using these coordinates. This is actually quite easy, at least for + and scalar product.

Take a vector **v**, and represent it as an arrow in the (x,y)-plane with its tail at the origin. The head of the vector is at some point (v_1, v_2). Furthermore, these two numbers, in order, tell us how to construct

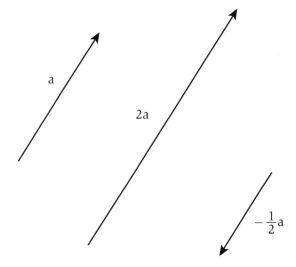

Figure 2-2. Scalar products.

the vector $\mathbf{v}$: it is the vector from the origin to the point (v_1, v_2). So, the numbers represent the vector uniquely. The numbers v_i are called the *components* of the vector $\mathbf{v}$.

Using the formula of Pythagoras, we see that if the vector $\mathbf{v}$ is represented by the ordered pair (v_1, v_2), its length must be $\sqrt{v_1^2 + v_2^2}$. The direction is a little more subtle. As long as v_1 is nonzero, we can say that the vector has "slope" v_2/v_1, but this breaks down when $v_1 = 0$. It is a little bit nicer to use polar coordinates and say that the direction is the polar angle θ equivalent to (v_1, v_2), although this begs the question a little bit. In fact, people sometimes just draw the vector, point at the arrow, and say "that way."

Polar Coordinates

A vector whose length is 1 is called a *unit vector*. You can construct a unit vector with the same direction as any nonzero vector $\mathbf{v}$ by taking the scalar product of $1/|\mathbf{v}|$ with $\mathbf{v}$.

Since we know the length of a unit vector, the polar coordinates of the point corresponding to the head of the arrow are completely determined by the angle θ alone. In fact, the head of such a vector (in the plane) is on the unit circle, so using the definitions of sine and cosine, we see that for every unit vector $\mathbf{u}$ there is a θ such that $\mathbf{u}$ can be written as

$$(\cos\theta, \sin\theta).$$

This means that the vector of length r in the direction of θ can be written as

$$r(\cos\theta, \sin\theta)$$

because it has the same direction as $\mathbf{u}$ and r times the magnitude. This still works when r is zero or even negative.

Spherical Coordinates

A vector in 3-space has three components, since it takes all three x, y, and z coordinates to pinpoint the head. We would normally write such a vector $\mathbf{v}$ as $\mathbf{v} = (v_1, v_2, v_3)$.

If **v** is a unit vector in space, its head is on the unit sphere given by $\rho = 1$ in spherical coordinates. Thus, it is specified by the two angles θ and ϕ. We can use the conversion from spherical to polar coordinates from Chapter 1 to determine the relationship between the components and the angles θ and ϕ:

$$(\sin \phi \cos \theta, \sin \phi \sin \theta, \cos \phi).$$

This doesn't look like it has length 1, so we had better check. Keep in mind that for any angle α, $\sin^2 \alpha + \cos^2 \alpha = 1$.

The squared length of the vector (using the standard length formula) is

$$
\begin{aligned}
(\sin \phi \cos \theta)^2 + (\sin \phi \sin \theta)^2 + \cos^2 \phi &= \sin^2 \phi \cos^2 \theta + \sin^2 \phi \sin^2 \theta + \cos^2 \phi \\
&= \sin^2 \phi (\sin^2 \theta + \cos^2 \theta) + \cos^2 \phi \\
&= \sin^2 \phi (1) + \cos^2 \phi \\
&= 1
\end{aligned}
$$

as expected.

The vector in the direction of θ and ϕ with length ρ is just

$$\rho(\sin \phi \cos \theta, \sin \phi \sin \theta, \cos \phi).$$

Cylindrical Coordinates

Sometimes it is useful to express a vector with reference to cylindrical coordinates. After all, if we have three ways of doing something, it's usually the case that at some point each of the three is the most convenient. If the head of the vector **v** is at the point with cylindrical coordinates r, θ, and z, the components are

$$(r \cos \theta, r \sin \theta, z).$$

Notice that the length properties don't translate as well in this coordinate system as they do in spherical coordinates. This is because cylindrical r is not a "pure" distance; it is the length of the projection of the vector onto the (x,y)-plane, and we know that $\tan \phi = r/z$, that is, $r = z \tan \phi$.

Vector Algebra in Coordinates

Vector algebra becomes very powerful when expressed in coordinates. We'll do addition in the plane, then jump to the more general case of vectors in space, which have 3 components (which is why space is often more precisely called 3-space). The most general situation is a vector with n components, where n is a whole number; such vectors live in n-space. Using this definition, a fourth dimension is rather unmystical and quite practical.

Suppose that $\mathbf{v} = (v_1, v_2)$ and $\mathbf{w} = (w_1, w_2)$ are vectors in the plane. As in Figure 2-3, we draw them and form the resultant $\mathbf{v} + \mathbf{w}$ in the usual way by placing the tail of one on the head of the other (and we know that it doesn't matter which order we do this in).

It seems obvious from the picture that the head of $\mathbf{v} + \mathbf{w}$ is at the point with coordinates $(v_1 + w_1, v_2 + w_2)$. You might want to check

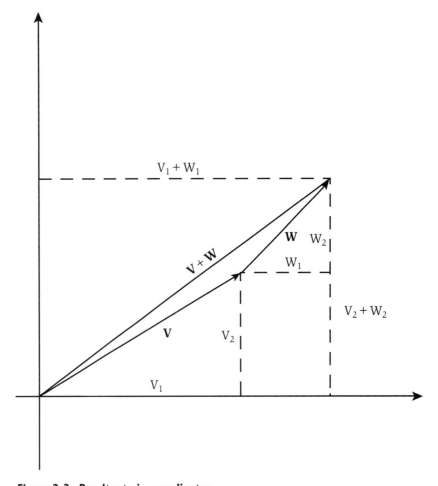

Figure 2-3. Resultants in coordinates.

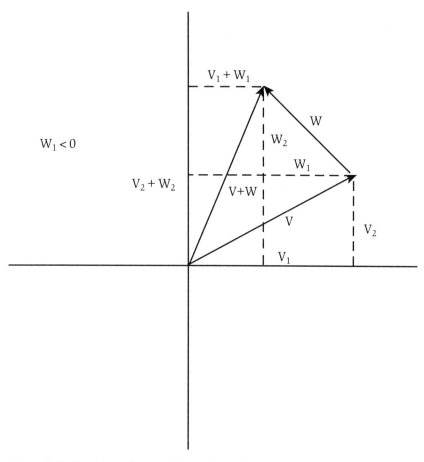

Figure 2-4. Resultants in coordinates (negative component).

that this is still the case when one of the components is negative, as in Figure 2-4.

What about scalar multiplication? What we want to do is change the length of the vector **v** without changing the direction. Since the slope v_2/v_1 (defined as long as $v_1 \neq 0$) must stay the same, what we have to do is multiply each component by the same factor. Hence

$$r(v_1, v_2) = (rv_1, rv_2).$$

Notice that $r(v_1, v_2)$ has the same slope, since $rv_2/rv_1 = v_2/v_1$, and the length is

$$|r(v_1, v_2)| = \sqrt{(rv_1)^2 + (rv_2)^2}$$
$$= \sqrt{r^2 v_1^2 + r^2 v_2^2}$$
$$= \sqrt{r^2(v_1^2 + v_2^2)}$$
$$= |r|\sqrt{v_1^2 + v_2^2}$$
$$= |r||(v_1, v_2)|$$

We have to take the absolute value of r in the last lines because when we are measuring distances, we always use the positive square root. Even if the original r had been negative, the square root of its square is positive.

In 3-space, we use the same basic form of these rules, but of course there is an extra component:

$$(v_1, v_2, v_3) + (w_1, w_2, w_3) = (v_1 + w_1, v_2 + w_2, v_3 + w_3) \quad \text{and}$$
$$r(v_1, v_2, v_3) = (rv_1, rv_2, rv_3).$$

Inner Product

The *inner product* (sometimes called the *dot product*) is the second of our multiplications. This product eats two vectors and produces a number. While this might be of some interest in theory, it works out that the number produced tells us a lot about the relative position of the vectors. In fact, dot product is one of the best ways we know to measure angles.

Here is the formal definition, in 2-space and in 3-space:

$$(v_1, v_2) \cdot (w_1, w_2) = v_1 w_1 + v_2 w_2$$
$$(v_1, v_2, v_3) \cdot (w_1, w_2, w_3) = v_1 w_1 + v_2 w_2 + v_3 w_3.$$

For example, $(2, 1, -1) \cdot (3, 2, 1) = 6 + 2 - 1 = 7$.

We call this a "product" because it satisfies some of the properties of ordinary multiplication. To prove these, you write out both sides of the equation in coordinates and manipulate the symbols until the two are equal. We don't refer to these very much, but in the interest of completeness, the properties are

$$\mathbf{v} \cdot \mathbf{w} = \mathbf{w} \cdot \mathbf{v} \quad \text{and}$$
$$\mathbf{v} \cdot (\mathbf{w}_1 + \mathbf{w}_2) = \mathbf{v} \cdot \mathbf{w}_1 + \mathbf{v} \cdot \mathbf{w}_2.$$

There is another property, that is, in a sense, also shared with ordinary multiplication, but the end result isn't really significant in that context, while for vectors it is very significant. Namely, what happens if you take the dot product of a vector with itself? You get its square, but in a transformed way:

$$\mathbf{v} \cdot \mathbf{v} = |\mathbf{v}|^2$$

Dot Product and Angles

In order to see the connection between dot product and angles, we need to take another look at the formula of Pythagoras for right triangles and make it a little more general. This is because not every triangle is a right triangle.

Consider the triangle in Figure 2-5 below, where, as usual, the side opposite the angle A has length a, etc. We are going to assume that we know b and c and the measure of the angle A.

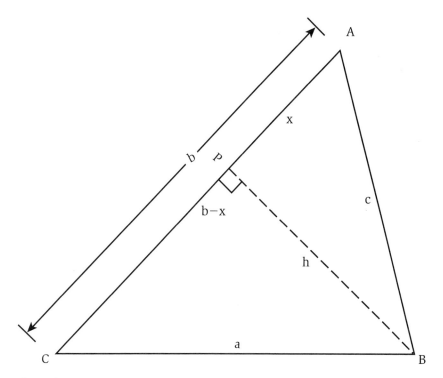

Figure 2-5. A generic triangle.

Drop a perpendicular from B to the side of length b, and call the foot of the perpendicular P. Let x be the length of $\overline{AP}$, so $b - x$ is the length of $\overline{PC}$.

Using trigonometry, we see that $x = c \cos A$.

Using Pythagoras, we see that $a^2 = h^2 + (b - x)^2$, and that $c^2 = h^2 + x^2$. Expanding and substituting for x, we get

$$
\begin{aligned}
a^2 &= h^2 + (b - x)^2 \\
&= h^2 + b^2 + x^2 - 2bx \\
&= b^2 + h^2 + x^2 - 2bx \\
&= b^2 + c^2 - 2bx \\
&= b^2 + c^2 - 2bc \cos A.
\end{aligned}
$$

That is, we can determine a if we know b, c, and A. (This is the side-angle-side version of triangle congruence, by the way.) This is called the *law of cosines*.

There might be some concern if P, the foot of the perpendicular, is not between A and C, as in Figure 2-6. In this case, x is bigger than b, so we should really use $x - b$ for the length of $\overline{PC}$, but since we square the length of $\overline{PC}$ in the derivation, the sign is not important.

The Law of Cosines

The law of cosines is a slightly more general way to relate lengths and angles.[5] This gives us another way to look at the dot product.

Start in Figure 2-7 with two nonzero vectors $\mathbf{v} = (v_1, v_2)$ and $\mathbf{w} = (w_1, w_2)$. Their dot product is $v_1 w_1 + v_2 w_2$.

Call the third side of the triangle above $\mathbf{s}$. It is clear that $\mathbf{v} + \mathbf{s} = \mathbf{w}$; so, subtracting, we know that $\mathbf{s} = \mathbf{w} - \mathbf{v}$. The three sides of the triangle have length $|\mathbf{w} - \mathbf{v}|$, $|\mathbf{v}|$, and $|\mathbf{w}|$, so we can determine the (acute) angle between $\mathbf{v}$ and $\mathbf{w}$ using the law of cosines. If θ is this angle, then we know

$$|\mathbf{w} - \mathbf{v}|^2 = |\mathbf{v}|^2 + |\mathbf{w}|^2 - 2|\mathbf{v}||\mathbf{w}| \cos \theta.$$

We know that for any vector, its length squared is its dot product with itself. There are three such terms in the formula above, so sub-

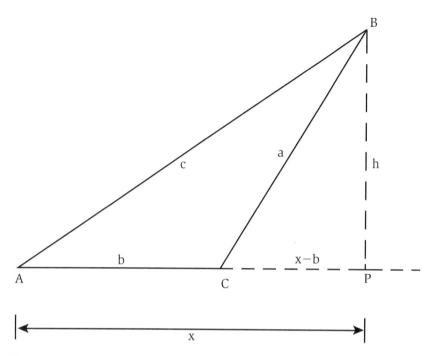

Figure 2-6. Another generic triangle.

stitute the dot product form of the length squared (this brings the dot product into play). The equality becomes

$$(\mathbf{w} - \mathbf{v}) \cdot (\mathbf{w} - \mathbf{v}) = \mathbf{v} \cdot \mathbf{v} + \mathbf{w} \cdot \mathbf{w} - 2|\mathbf{v}||\mathbf{w}| \cos \theta.$$

This is a much easier calculation because we can write out the three dot products in components. The two length factors seem a little annoying, so we will pull a common mathematician's trick and leave them as is for now instead of trying to simplify them. We get

$$(w_1 - v_1)^2 + (w_2 - v_2)^2 = v_1^2 + v_2^2 + w_1^2 + w_2^2 - 2|\mathbf{v}||\mathbf{w}| \cos \theta.$$

We can multiply out the squares on the left. This leads to a nice situation: each side of the equation has the terms v_1^2, v_2^2, w_1^2, and w_2^2, so we eliminate these and get

$$-2v_1 w_1 - 2v_2 w_2 = -2|\mathbf{v}||\mathbf{w}| \cos \theta \text{ or}$$
$$-2(v_1 w_1 + v_2 w_2) = -2|\mathbf{v}||\mathbf{w}| \cos \theta.$$

Dividing each side of the equation by -2 leaves us with

$$v_1 w_1 + v_2 w_2 = |\mathbf{v}||\mathbf{w}|\cos\theta$$

that is,

$$\mathbf{v} \cdot \mathbf{w} = |\mathbf{v}||\mathbf{w}|\cos\theta$$

where θ is the acute angle between them.

Application: Distances Along a Sphere

Let's look at an important example that illustrates the use of the dot product to measure angles. Given any two points P_1 and P_2 on a sphere of radius r, find the great-circle distance between them. (This will help solve the problem of finding the great-circle distance between any two points along the Earth's surface.)

The keys to this calculation are radian measure and the dot product. The reason is that the plane formed by the two points in question and the center of the sphere cuts the sphere in a great circle whose radius is r (this is the definition of *great circle*, you'll recall). By taking the dot product of the two displacement vectors, we can determine the radian measure of the angle α at the center of the

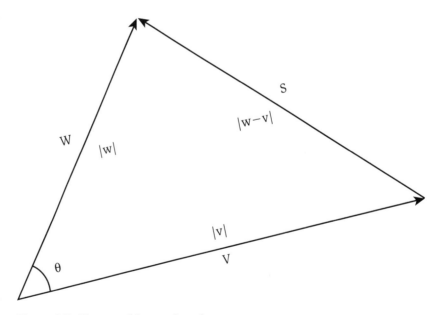

Figure 2-7. Vectors of known length.

wedge, and, knowing this and the radius, we can determine the length of the circular arc.

To find the dot product of the two displacement vectors, we need to do some preliminary work. First, we have to determine coordinates. Since we are working with a sphere, it seems most convenient to use spherical coordinates. But to do this we need to determine an origin and x, y, and z axes, because the spherical distance ρ is distance from the origin, and the spherical angles θ and ϕ are measured with reference to the axes. Now, without two particular points in mind, we do not have any preferred candidates for the axes (if one of the points were fixed it would probably be best to make it a pole), so we will just assume that the axes have been fixed. The final result is independent of the choice of axes.

Put the origin O at the center of the sphere, and suppose that the spherical angles describing each P_i are θ_i and ϕ_i (i = 1 or 2). We can assume that these are known. Of course, $\rho = r$. See Figure 2-8.

There are two displacement vectors $\mathbf{v}_i = \overline{OP_i}$, and each of these has components

$$r(\sin \phi_i \sin \theta_i, \sin \phi_i \cos \theta_i, \cos \phi_i).$$

Each of these has length r.

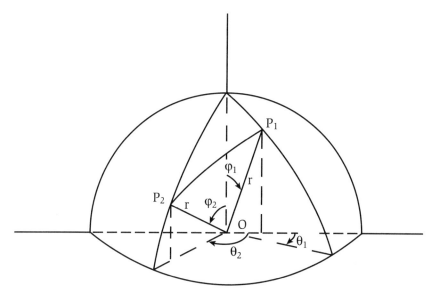

Figure 2-8. Two points on a sphere.

We can calculate the dot product $\mathbf{v}_1 \cdot \mathbf{v}_2$ two different ways, once in coordinates and then using the law of cosines, and the two results have to be equal.

Using the law of cosines,

$$\mathbf{v}_1 \cdot \mathbf{v}_2 = |\mathbf{v}_1||\mathbf{v}_2| \cos \alpha$$
$$= r^2 \cos \alpha.$$

From here, it is a simple matter to solve for $\cos \alpha$, and, from there, to solve for α. In fact,

$$\cos \alpha = \frac{1}{r^2} \ \mathbf{v}_1 \cdot \mathbf{v}_2 \quad \text{so}$$
$$\alpha = \arccos(\frac{1}{r^2} \ \mathbf{v}_1 \cdot \mathbf{v}_2).$$

The messy part of this calculation is finding $\mathbf{v}_1 \cdot \mathbf{v}_2$ in coordinates. The general formula is long, complicated, and completely unilluminating. In practice, you use a calculator to determine it. But even with a calculator it isn't easy. Here's how to start. You already know the angles θ_i and ϕ_i, so you get a calculator and determine the values of $\sin \phi_i$, $\cos \phi_i$, $\sin \theta_i$, and $\cos \theta_i$ to as many decimal places as you feel necessary. That's eight calculations, in case you are keeping count. Then you determine the products $\sin \phi_i \sin \theta_i$, and $\sin \phi_i \cos \theta_i$. That's four more calculations, bringing the total to twelve.

Once you have done these twelve calculations, you have determined the three components

$$(\sin \phi_i \cos \theta_i, \sin \phi_i \sin \theta_i, \cos \phi_i).$$

You have to multiply by r to get the (x,y,z) coordinates of the points P_i. Now you can take the dot product!

But wait: note that when you take the dot product in coordinates, each term of the result has a factor of r^2. The geometric version of the dot product was just $r^2 \cos \alpha$, so you can ignore r altogether (at least until you want to convert from radians to length units) by dividing out by this common factor. The geometry of this result is that the angle between the vectors doesn't depend on the radius at all; points on concentric spheres would have the same angle, although the distance between them would vary with the radius. See Figure 2-9.

For example, look at the points on the unit sphere P with $\phi = 36.7°$, $\theta = 60.4°$ and Q with $\phi = 26.0°$, $\theta = 22.7°$. (These points correspond to the latitude and longitude of Goose Bay, Labrador, which is at approximate latitude N53.3°, and Keflavik, Iceland, which is at approximate latitude N64.0°.) The Cartesian coordinates of P are $(0.5196, 0.2952, 0.8018)$ and those of Q are $(0.1692 , 0.4044, 0.8988)$; you can check that each of these is on the unit sphere since the sum of the squares is very close to 1.

The dot product is $0.0879 + 0.1194 + 0.7207 = .9280$, so the angle α formed by P, the center, and Q is $\arccos(0.9280) \approx 21.88°$. If the Earth were a perfect sphere of circumference 21,600 NM, the distance would be 1313 NM; according to WGS–84, at these latitudes each degree is about 60.153 NM (taking the length of a degree at the mid-latitude), so the distance is about 1316 NM.

Special Case: Points at the Same Latitude

One special case of this is especially useful: given two points with the same ϕ, what is the distance between them? This would be the way to determine the distance to a point due east or due west of a known

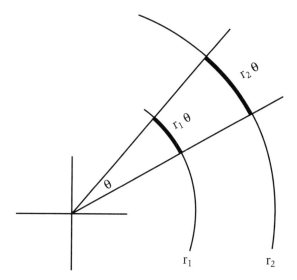

Figure 2-9. Projections of points on concentric spheres.

point on the Earth. We'll only do the case of sphere of radius 1; you need to multiply the end result of this calculation by the actual radius in practice.

We can assume that one of the points is on the great circle where $\theta = 0$, so the coordinates of one such point are $(0, \sin\phi, \cos\phi)$, while the other's coordinates are

$$(\sin\phi\cos\theta, \sin\phi\sin\theta, \cos\phi).$$

Thus, the cosine of the distance is $\sin^2\phi\cos\theta + \cos^2\phi$, which is the dot product of the two vectors.

There is a useful approximation of this that is valid when θ—the difference between the longitudes—is small. The method here is to estimate the great-circle route between the points using the small circle of points with the given value of ϕ. See Figure 2-10.

This circle has radius $\cos\phi$, so its circumference is $2\pi\cos\phi$. In par-

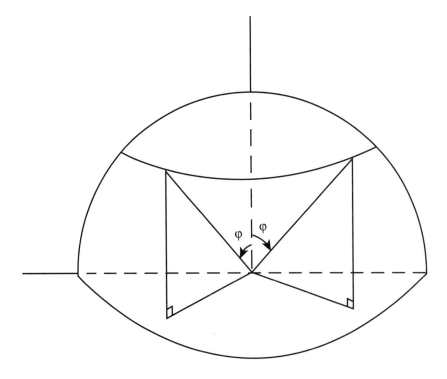

Figure 2-10. Small circle at latitude ϕ.

ticular, two points $1°$ apart are separated by about $2\pi \cos\phi/360$ distance units.

This is even easier on the Earth's sphere if the units are nautical miles. One degree of latitude is, by definition, 60 nautical miles of arc. When the latitude is ϕ, therefore, $1°$ of longitude is $60 \cos\phi$ nautical miles.

Vectors and Lines and Planes

Now that we have a new way to measure angles (the dot product), it's natural to see if there are any interesting angles to measure. Navigation is about latitude and longitude (or, what's almost the same thing, ϕ and θ), and these are angles. Let's take another look at geodetic latitude, L_N.

The geodetic latitude (Figure 1-43 in Chapter 1) of a point P is measured with reference to the normal to Earth's tangent plane at P; L_N is the angle between this normal and the line $\overline{OE}$ from the center of the Earth O to the point E where the meridian of P meets the equator.

We can interpret each of these lines—the normal and $\overline{OE}$—as a vector. The latter becomes just a displacement vector called, say, $\mathbf{E}$, but the former has no intrinsic length. We can take the length to be 1 and call this *unit normal* $\mathbf{N}$. In fact, using the law of cosines and remembering that $|\mathbf{N}| = 1$,

$$\mathbf{N} \cdot \mathbf{E} = |\mathbf{N}||\mathbf{E}| \cos(\mathbf{L_N})$$
$$= R \cos(L_N),$$

where R is the radius of the Earth.

The problem is to get our hands on the normal vector $\mathbf{N}$. This is normal to a plane; remember that the normal direction is perpendicular to all of the lines in the plane. We can rephrase this now: The normal vector $\mathbf{N}$ is perpendicular to every vector $\mathbf{v}$ in the plane. Since $\cos\pi/2 = 0$, this means that

$$\mathbf{N} \cdot \mathbf{v} = 0$$

for any vector $\mathbf{v}$ in the plane.

We can use this concept to translate back-and-forth between equations for planes and geometric descriptions. The same ideas work for lines in a fixed plane, too, and since this is a little easier to visualize, let's look at that case first.

Equations of Lines

The most common way to describe a line by an equation is the *slope-intercept* form. Remember that the slope of a line is the ratio "rise over run," that is, how many steps up you take in response to each step to the right. See Figure 2-11.

For some reason, the letter m has come to be associated with slope. The slope of a vertical line is not defined, since the run is always zero.

But the slope alone doesn't completely determine a line, because any parallel line has the same slope. We need a point on the line to pin it down.

Any nonvertical line meets the y axis at a point called the y-intercept, and this is what we use for the slope-intercept form. The coordinates of this point have the form $(0,b)$. Let (x,y) be an arbitrary point on the line. Then using the definition of slope,

$$\frac{y - b}{x - 0} = m$$

so, solving for y,

$$y = mx + b.$$

This form of the equation—solving for y—is very useful in calcu-

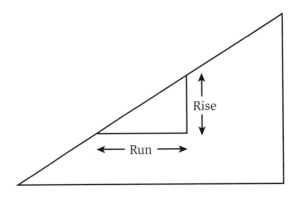

Figure 2-11. Slope.

lus, and that's why it is so widely taught. But this form doesn't work so well for planes, or even for some applications involving lines.

So, choose an *arbitrary* point (a,b) on a line of slope m, and again let (x,y) denote a variable point on the line. Using the definition of slope again,

$$\frac{y-b}{x-a} = m$$

and the only simplification worth making here is to clear the denominator and write

$$y - b = m(x - a).$$

This is the *point-slope* form of the equation of a line.

Parametric Description of a Line

In many cases, it is easier to study a mathematical object in terms of a *parameter* or parameters. One example of this is the unit circle in the plane. You can specify points by their x and y coordinates, like $(0.8660, 0.5)$ or by the corresponding angle θ, in this case $30°$. The whole circle can be described by the single *parameter* θ.

Another example is the curve given by the equation $y^2 = x^3 + x^2$. It is difficult to find the coordinates of points on this curve, but it can be parametrized with a parameter t such that $x = t^2 - 1$ and $y = t^3 - t$; you can check this by calculating $y^2 = (t^3 - t)^2$ and comparing it with $x^3 + x^2 = (t^2 - 1)^3 + (t^2 - 1)^2$. The parametrized version is much easier to work with.

We can use vectors to parametrize a line, and this parametrization has the two advantages of being easy to generalize to arbitrary dimension and of clarifying the role of normals.

To parametrize a line, we need to specify its direction as a vector rather than as a slope. And of course we need a point on the line, which can also be specified as a vector (the displacement from the origin to the point). Here we run into a little ambiguity: If we have both vectors and points, all of which are described by ordered pairs (or triples, in 3-space), how do we tell which are vectors and which are numbers? The answer is that it doesn't matter; interpret what you see in the most useful way.

Let $\mathbf{v} = (v_1, v_2)$ be a vector in the direction of the line of interest, and let $\mathbf{p} = (p_1, p_2)$ be the point on the line. If the direction is specified by a slope m, we can use $(1, m)$ for $\mathbf{v}$, but we want to develop other ideas about lines, so we will just take (v_1, v_2) as given.

How can we get to an arbitrary point $\mathbf{X} = (x, y)$ on this line? First, we can follow the displacement vector to the point $\mathbf{p} = (p_1, p_2)$. Then we can move parallel to $\mathbf{v}$ for some distance to get to (x, y). The displacement vector from $\mathbf{p}$ to $\mathbf{X}$ is parallel to $\mathbf{v}$, so for some scalar t,

$$\mathbf{X} - \mathbf{p} = t\mathbf{v}.$$

We can solve for the unknown (namely, $\mathbf{X}$), and write

$$\mathbf{X} = \mathbf{p} + t\mathbf{v}.$$

Now t is the parameter. See Figure 2-12.

The nice thing is that if $\mathbf{v}$ and $\mathbf{p}$ are vectors in 3-space, then the exact same equation describes a line in 3-space, namely, the line through $\mathbf{p}$ parallel to $\mathbf{v}$. This is the nice generalization alluded to above.

Equations Using the Normal

Often you are given a normal to the line of interest: describe the line through the point $\mathbf{p}$ *perpendicular* to $\mathbf{n}$. At least in the (x, y)-plane,

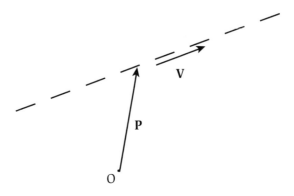

Figure 2-12. Parametrized line.

knowing the perpendicular (or normal) is at least as good as knowing the slope. This happens in the reduction of celestial observations in Chapter 3.

Again, let $\mathbf{X}$ denote a variable point on the line. As before, the displacement vector $\mathbf{X} - \mathbf{p}$ is parallel to the line, so it is therefore perpendicular to the normal $\mathbf{n}$. But two nonzero vectors are perpendicular if and only if their dot product is zero! Thus we have

$$(\mathbf{X} - \mathbf{p}) \cdot \mathbf{n} = 0$$

or

$$\mathbf{X} \cdot \mathbf{n} = \mathbf{p} \cdot \mathbf{n}.$$

The thing is that $\mathbf{p} \cdot \mathbf{n}$ is a constant. Let $c = \mathbf{p} \cdot \mathbf{n}$, and suppose that $\mathbf{n} = (a, b)$ in coordinates. As usual, write $\mathbf{X} = (x, y)$ in coordinates. Now we have that $\mathbf{X} \cdot \mathbf{n} = ax + by$, so the equation becomes

$$ax + by = c.$$

When the equation has this form, we can find a normal vector by inspection. For example, $(3, -2)$ is a normal vector for the line given by $3x - 2y = 1$.

This generalizes directly to equations of planes in 3-space. We'll need these later in order to measure azimuths, that is, the magnetic or true direction between points on the Earth's surface. Consider the plane through $\mathbf{p}$ normal to $\mathbf{n} = (a, b, c)$. Let d denote $\mathbf{p} \cdot \mathbf{n}$, as before. It is a constant for this plane. Now, if $\mathbf{X} = (x, y, z)$ is a variable point on the plane, then

$$(\mathbf{X} - \mathbf{p}) \cdot \mathbf{n} = 0$$

or

$$\mathbf{X} \cdot \mathbf{n} = \mathbf{p} \cdot \mathbf{n}.$$

These are the exact equations we used for lines in the plane. Now we bring the coordinates back into the picture and get the equation

$$ax + by + cz = d.$$

And now, again, the normal is easy to read: if the plane has equation $ax + by + cz = d$, then (a, b, c) is a normal vector.

The Cross Product

In the above discussion, we were given the normal vector to the plane, and we used it to find an equation for the plane. This makes some sense in context because the most interesting plane for us is the horizon (the tangent plane to the Earth at our feet), and we can take the zenith direction as the normal. But what about the old adage that "three points determine a plane"? Shouldn't we be able to take three points $\mathbf{X}_0$, $\mathbf{X}_1$, and $\mathbf{X}_2$ in space and determine an equation for the plane they define? This is the way the planes used in celestial navigation are defined: what is the plane through Kennedy airport in New York, the center of the Earth, and Heathrow airport outside of London? One leg of its intersection with the surface of the Earth is the great-circle route from New York to London.

In Figure 2-13, suppose that $\mathbf{N}$ is a normal vector to the plane these points define. Right now, $\mathbf{N}$ is the unknown. Since all $\mathbf{X}_i$ are in the plane, their differences are in the plane, so $\mathbf{V}_1 = \mathbf{X}_1 - \mathbf{X}_0$ and $\mathbf{V}_2 = \mathbf{X}_2 - \mathbf{X}_0$ are vectors in our plane. This means that each of these is perpendicular to $\mathbf{N}$, so

$$(\mathbf{X}_1 - \mathbf{X}_0) \cdot \mathbf{N} = 0 \text{ and}$$
$$(\mathbf{X}_2 - \mathbf{X}_0) \cdot \mathbf{N} = 0,$$

that is,

$$\mathbf{V}_1 \cdot \mathbf{N} = 0 \text{ and}$$
$$\mathbf{V}_2 \cdot \mathbf{N} = 0.$$

In these equations, the $\mathbf{V}_i$ are known, and $\mathbf{N}$ is the unknown.

Introducing coordinates $\mathbf{V}_i = (x_i, y_i, z_i)$ and $\mathbf{N} = (n_1, n_2, n_3)$, these equations become

$$x_1 n_1 + y_1 n_2 + z_1 n_3 = 0 \text{ and}$$
$$x_2 n_1 + y_2 n_2 + z_2 n_3 = 0.$$

We can easily solve for two of the n_i, but without a third equation we can't find all of them. The problem here is that we know the direction of $\mathbf{N}$ but not its length. Once we have a length, we still need to decide whether $\mathbf{N}$ points "up" or "down."

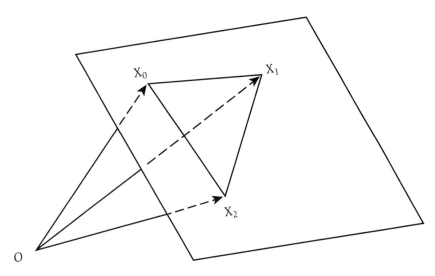

Figure 2-13. Three points determine a plane.

Normalizing the Normal

The first impulse is to make $\mathbf{N}$ a unit vector, but this leads to an incredible mess. Instead, we should mimic the great success of the dot product and try to have $|\mathbf{N}|$ encode some aspect of the relative position of the two vectors $\mathbf{V}_i$.[6]

There are many candidates for a number describing the relative position of the two vectors. To simplify the search, we are going to look at some "special" vectors $\mathbf{V}_i$; namely, suppose that each $\mathbf{V}_i$ has the form $\mathbf{V}_i = (x_i, y_i, 0)$. With this choice, we know that $\mathbf{N}$ has to point along the z axis, and it therefore has the form $(0, 0, n_3)$.

The dot product told us the angle θ between the two vectors, but the other factor was their length: $\mathbf{V}_1 \cdot \mathbf{V}_2 = |\mathbf{V}_1||\mathbf{V}_2| \cos \theta$. Keeping the angle but changing the lengths changes the dot product. But keeping the angle and changing the lengths also changes something else, namely the area of the parallelogram determined by the two vectors; see Figure 2-14.

Is there a nice way to express the area of the parallelogram determined by two nonzero vectors $\mathbf{V}_1$ and $\mathbf{V}_2$ when the angle between them is θ? We can determine this area by comparing it with the area of a rectangle whose base is $\mathbf{V}_1$ and whose height is determined by

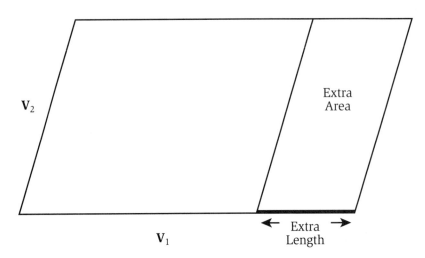

Figure 2-14. Changing lengths and changing areas.

the length of the perpendicular to $\mathbf{V}_1$ through the end of $\mathbf{V}_2$, as in Figure 2-15.

Call the foot of this perpendicular $\mathbf{F}$ and let h be its length. The rectangle in question has area $h|\mathbf{V}_1|$. Extending the two sides parallel to $\mathbf{V}_1$ and dropping perpendiculars from the corners, we see that the rectangle includes a triangle congruent to $O\mathbf{F}\mathbf{V}_2$ but also excludes a triangle congruent to $O\mathbf{F}\mathbf{V}_2$; this means that the area of the parallelogram is the same as that of the rectangle, namely,

$$h|\mathbf{V}_1|.$$

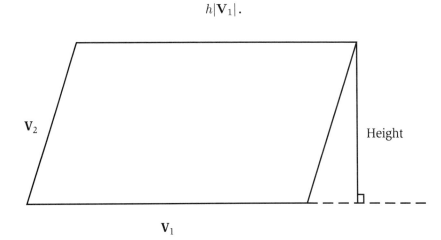

Figure 2-15. Height of a parallelogram.

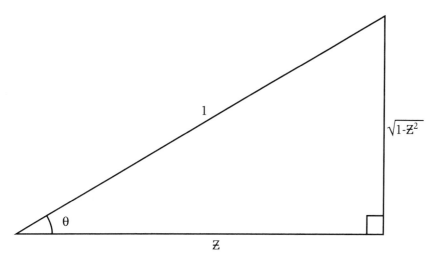

Figure 2-16. Angle with given cosine.

We use trigonometry to determine h: by definition, $\sin\theta = h/|\mathbf{V}_2|$, so

$$h = |\mathbf{V}_2|\sin\theta$$

and the area is

$$|\mathbf{V}_1||\mathbf{V}_2|\sin\theta.$$

This is the magnitude of $\mathbf{V}_1 \times \mathbf{V}_2$.

The Cross Product in Coordinates

What does the cross product look like in coordinates?

The first problem is that we know $\cos\theta$ instead of θ, so how can we get $\sin\theta$? The technique is to skip the intermediate step and go back to the definition of $\cos\theta$. Suppose that $z = \cos\theta$; this means that θ fits into a right triangle whose adjacent side is z and whose hypotenuse is 1; this is shown in Figure 2-16.

Since this is a right triangle, we know (by Pythagoras) that the opposite side is $\sqrt{1-z^2}$, and this means that $\sin\theta$ is also $\sqrt{1-z^2}$.

Introduce the coordinates $\mathbf{V}_i = (x_i, y_i)$. We know from the dot product that

$$\cos\theta = (x_1x_2 + y_1y_2)/|\mathbf{V}_1||\mathbf{V}_2|.$$

We need $|\mathbf{V}_1||\mathbf{V}_2|\sin\theta$. Expanding,

$$|\mathbf{V}_1||\mathbf{V}_2|\sin\theta = |\mathbf{V}_1||\mathbf{V}_2|\sqrt{1 - (\frac{x_1x_2 + y_1y_2}{|\mathbf{V}_1||\mathbf{V}_2|})^2}$$
$$= \sqrt{|\mathbf{V}_1|^2|\mathbf{V}_2|^2 - (x_1x_2 + y_1y_2)^2}.$$

The length terms become squared when they move inside the square root sign.

This is much easier to deal with, since $|\mathbf{V}_1|^2$ is just $\mathbf{V}_i \cdot \mathbf{V}_i$. Expanding, we see

$$|\mathbf{V}_1||\mathbf{V}_2|\sin\theta = \sqrt{(x_1^2 + y_1^2)(x_2^2 + y_2^2) - (x_1x_2 + y_1y_2)^2}$$
$$= \sqrt{x_1^2x_2^2 + x_1^2y_2^2 + y_1^2x_2^2 + y_1^2y_2^2 - x_1^2x_2^2 - 2x_1x_2y_1y_2 - y_1^2y_2^2}$$
$$= \sqrt{x_1^2y_2^2 - 2x_1x_2y_1y_2 + y_1^2x_2^2}$$
$$= \sqrt{(x_1y_2 - y_1x_2)^2}.$$

This almost completes the picture, but there is a sign problem in taking the square root of the square. Remember that we have the option of reversing the direction of the normal vector while keeping its length, that is, we can negate the normal vector. In this case, we will decide to use another *right-hand rule*: point the fingers of your right hand from $\mathbf{V}_1$ to $\mathbf{V}_2$, and make the normal point in the direction of your right thumb, (see Figure 2-17).

If the vectors $\mathbf{V}_i$ have "positive orientation" (that is, the acute angle between them is formed by starting at $\mathbf{V}_1$ and moving counterclockwise), then n_3 must be positive. Hence

$$(x_1, y_1, 0) \times (x_2, y_2, 0) = (0, 0, x_1y_2 - y_2x_1).$$

By symmetry (with some care to signs), you can work out that

$$(x_1, 0, z_1) \times (x_2, 0, z_2) = (0, x_2z_1 - x_1z_2, 0) \text{ and}$$
$$(0, y_1, z_1) \times (0, y_2, z_2) = (y_1z_2 - z_1y_2, 0, 0).$$

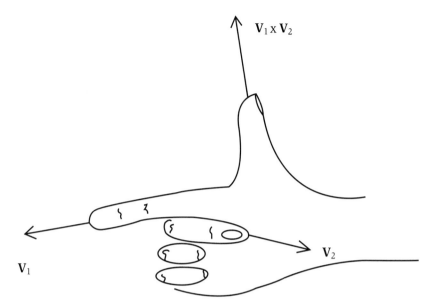

Figure 2-17. Right-hand rule.

Calculating Cross Products

What about the general form of the cross product? This is easier if we introduce some notation used by physicists for the unit vectors along the axes (Figure 2-18):

$$\mathbf{i} = (1, 0, 0)$$
$$\mathbf{j} = (0, 1, 0)$$
$$\mathbf{k} = (0, 0, 1).$$

Using the right-hand rule and the fact that all of the parallelograms generated by pairs of these three are squares of side one, we see that

$$\mathbf{i} \times \mathbf{j} = \mathbf{k}$$
$$\mathbf{j} \times \mathbf{k} = \mathbf{i} \text{ and}$$
$$\mathbf{k} \times \mathbf{i} = \mathbf{j}.$$

Notice that any vector (a, b, c) can be written as $a\mathbf{i} + b\mathbf{j} + c\mathbf{k}$.

A *matrix* is a rectangular array of numbers; the *determinant* of a 2×2 (that is, 2 rows and 2 columns) matrix $\begin{pmatrix} x_1 & y_1 \\ x_2 & y_2 \end{pmatrix}$ is defined to be $x_1 y_2 - x_2 y_1$, which is exactly the z-component of

$(x_1, y_1, 0) \times (x_2, y_2, 0)$.[7] The determinant of an $n \times n$ matrix is defined inductively. First if M is the matrix whose entries are m_{ij}, then use M_{ij}[8] to denote the $(n-1) \times (n-1)$ matrix formed by removing row i and column j from M. Then the determinant of M is calculated by choosing a row i and finding

$$(-1)^{i+1} m_{i1} \det M_{i1} + (-1)^{i+2} m_{i2} \det M_{i2} + \cdots + (-1)^{i+n} m_{in} \det M_{in}.$$

Luckily, we need to do this only for the 3×3 case:

$$\det \begin{pmatrix} a_1 & a_2 & a_3 \\ b_1 & b_2 & b_3 \\ c_1 & c_2 & c_3 \end{pmatrix}$$

$$= a_1 \det M_{11} - a_2 \det M_{12} + a_3 \det M_{13}$$

$$= a_1 \det \begin{pmatrix} b_2 & b_3 \\ c_2 & c_3 \end{pmatrix} - a_2 \det \begin{pmatrix} b_1 & b_3 \\ c_1 & c_3 \end{pmatrix} + a_3 \det \begin{pmatrix} b_1 & b_2 \\ c_1 & c_2 \end{pmatrix}.$$

Using the formal concept of the determinant function we can write out a "formula" for the general cross product. It is:

$$(x_1, y_1, z_1) \times (x_2, y_2, z_2) = \det \begin{pmatrix} \mathbf{i} & \mathbf{j} & \mathbf{k} \\ x_1 & y_1 & z_1 \\ x_2 & y_2 & z_2 \end{pmatrix}.$$

It is possible, although difficult, to make rigorous sense of this as a determinant calculation; whether one does so or not, it is perfectly OK to use it as a way of organizing the calculation. Luckily, we don't need to do too many of these.

Examples of Planes

To cement this all together, let's find an equation for the plane through $\mathbf{p_0} = (1, 1, 1)$, $\mathbf{p_1} = (2, 0, 3)$, and $\mathbf{p_2} = (-1, 2, 0)$. This kind of calculation is also needed for some calculations converting distance and bearing to latitude and longitude.

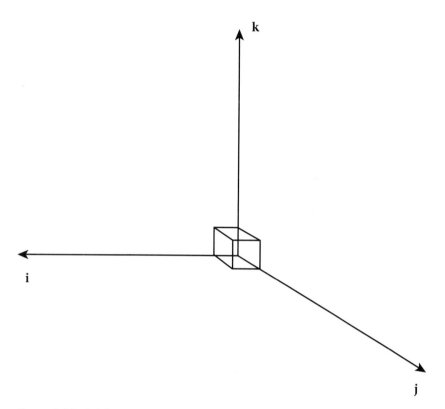

Figure 2-18. i, j, k.

Define $\mathbf{V}_i = \mathbf{p}_i - \mathbf{p}_0$; this makes $\mathbf{p}_0$ a kind of base point, and this choice is arbitrary; you would get an equivalent equation if you used one of the other $\mathbf{p}_i$ as the base.

Calculating, $\mathbf{V}_1 = (1, -1, 2)$ and $\mathbf{V}_2 = (-2, 1, -1)$, so a normal vector $\mathbf{n}$ is given by $\mathbf{V}_1 \times \mathbf{V}_2$, or

$$\mathbf{n} = \begin{pmatrix} \mathbf{i} & \mathbf{j} & \mathbf{k} \\ 1 & -1 & 2 \\ -2 & 1 & -1 \end{pmatrix}.$$

This is

$$\mathbf{i}\det \begin{pmatrix} -1 & 2 \\ 1 & -1 \end{pmatrix} - \mathbf{j}\det \begin{pmatrix} 1 & 2 \\ -2 & -1 \end{pmatrix} + \mathbf{k}\det \begin{pmatrix} 1 & -1 \\ -2 & 1 \end{pmatrix} =$$

$$= \mathbf{i}(1-2) - \mathbf{j}(-1+4) + \mathbf{k}(1-2)$$
$$= -\mathbf{i} - 3\mathbf{j} - \mathbf{k}$$
$$= (-1, -3, -1).$$

The first thing to notice is that $\mathbf{n} \cdot \mathbf{V}_1 = -1 + 3 - 2 = 0$, and that $\mathbf{n} \cdot \mathbf{V}_2 = 2 - 3 + 1 = 0$, so $\mathbf{n}$ really is perpendicular to the vectors in the plane.

Now what about the equation? If $\mathbf{X} = (x, y, z)$ is a variable point in the plane, then $\mathbf{X} - \mathbf{p_0}$ is a vector in the plane, so $\mathbf{n} \cdot (\mathbf{X} - \mathbf{p_0})$ must be zero. Calculating,

$$0 = \mathbf{n} \bullet (\mathbf{X} - \mathbf{p_0})$$
$$= (-1, -3, -1) \bullet (x - 1, y - 1, z - 1)$$
$$= -x + 1 - 3y + 3 - z + 1$$

so

$$x + 3y + z = 5$$

You can check that all three of the $\mathbf{p}_i$ satisfy this equation.

NOTES

1. The measurement of airspeed is surprisingly complicated and poorly understood, even by pilots. The distinction is unimportant in this example so we'll just let it pass. But see Chapter 5.

2. You are hereby challenged to write another true sentence involving those four cities!

3. At an appropriate scale, if necessary.

4. The word *represent* defines the difference between mathematicians and the rest of the world. Mathematicians work with representations of the world all the time, and non-mathematicians seem to confuse the representation with the reality.

5. In a sense, the formula of Pythagoras relates lengths and angles, but only in the case of a *right* angle.

6. Notice that the vectors $\mathbf{V}_i$ now seem to have taken over from the original Z_i.

7. Those who have studied vector calculus might recall that the determinant appears in the multivariable change-of-variables formula for integrals, exactly because it is the area factor.

8. Note the use of a capital letter.

CHAPTER 3

Navigating by the Stars

Introduction

The dot and cross products are the only mathematical tools one needs to do *celestial navigation*.[1] Forms of celestial navigation were used by Columbus in the fifteenth century, by the Apollo astronauts (Farmer), and are still in use today. The U.S. Navy uses it (Kaplan).

One needs much more than mathematical tools to navigate. One must make measurements, which are based on knowledge of the stars and planets, which is based on the making of measurements, which has been refined by knowledge of the stars and planets, and so on *ad infinitum*. Terrible mistakes have been made by neglect of the theory or lack of care in the practice, and some continue to be made to this day.[2]

Measurements require instruments; instruments are manufactured and calibrated, and this requires knowledge of metallurgy, chemistry, optics, and more. Such knowledge is gained in the laboratory and the field as well as in the mathematician's study, and the neglect of any of these can lead to mistakes. This is not to say that every navigator needs to know the coefficient of expansion of every metal in every instrument, or the vibrational frequency of cesium atoms, or the chemical makeup of the upper atmosphere. Nor, to be honest, does every navigator need to know the law of cosines. But somebody somewhere has to know each of these things, and somebody some-where has to verify this knowledge and put the pieces together to design a protocol for finding where you are and how to get where you want to go.

In this chapter, you will learn how to navigate by the stars and planets. This knowledge applies directly to the problems of great-cir-cle navigation. You will also see how to make rough calculations of heading and bearing.

Although you may never use celestial navigation, its study will help you better visualize the problems of long-range navigation. And, it has given us two very important concepts that navigators use every day: the concept of line of position, and the formulas for distance and bearing between two points. The sextant may fall into disuse, but the mathematics of celestial navigation remains fundamental to all long-range navigation.

A Little History

This is not the place to discuss the history of celestial navigation. (Bowditch) presents pages and pages of methods and discusses their relative utility, and (Cotter) is a comprehensive historical treatise. However, one historical note is important because of the way it illus-trates the process of technological development.

Roughly speaking, one can specify a spot on the Earth's surface by giving its latitude and longitude. I say "roughly" because, as we have seen, there are many ways of measuring latitude, and surveyors have historically had a difficult time pinpointing longitudes of places not connected by land. A survey starts at a definite point called the *datum*, and all latitudes and longitudes are measured (as accurately as possible) with reference to the datum. The measurements are *dis-tances*: a surveyor can measure, say, one nautical mile from the datum in a specific direction along the surface. To convert the distance to lat-itude and longitude, the surveyor needs to use a geoid—a model of the Earth's surface. The choice of a different geoid changes the coor-dinates of points.

This isn't a problem when you want to use the coordinates to build a bridge across a river where both banks have been charted with respect to the same datum, but it is a problem when you want to send a guided missile across the sea to a point whose coordinates are determined with respect to a different datum. And it's a problem in long-range navigation today because countries use different geoids and datums. Navigators, especially electronic ones, refer to the datum of the home country. For most of the world this is a datum called WGS-84, described in Chapter 4. But some countries use a different datum, and the coordinates in the other datum can differ. In Australia I was (thankfully) warned to set my GPS to use the Australian datum before flying to an island in Great Barrier Reef.[3]

The inability to accurately determine longitude after a sea voyage was a major problem during the era of Pacific exploration. Navigators at that time would make a good guess at the distance made good and "convert" to longitude. These calculations were helped by the fact that voyages of exploration (such as Bougainville's) generally traveled due west across the Pacific, so the latitude was known, as was an approximate radius of the Earth, but they were hindered because the ships were traveling through unknown currents, making the navigator's estimates of distance made good either too big or too small. On the other hand, this was Eratosthenes' method; he used "distance made good" to approximate the difference in latitude between Syene and Alexandria. The method is inevitably inaccurate: It wasn't until World War II that the relationship between some islands in the well-traveled Mediterranean Sea was precisely determined using radar.[4]

Those questing for coordinates were misled by the relative ease of estimating latitude. There are many methods: latitude by Polaris (the North Star) was discussed in Chapter 1, but the Sun can also be used to determine latitude if you have previously determined how its apparent position moves across the Earth's surface over the course of the year (more on this later). This is why the "noon sight" was so important to ships at sea: the altitude of the Sun at "local apparent noon" (when the Sun crosses the observer's meridian; that is, when the Sun is at its highest point above the horizon for the day) gives an easy method to determine latitude.

If there are two coordinates, and you have an easy time determining one of them, it seems quite natural to put a lot of effort into determining the other. The quest for the "longitude" was one of the major intellectual problems from the sixteenth through nineteenth centuries, and it is widely alluded to in literature, both scientific and otherwise.[5]

This situation persisted for years and years, until Captain T. H. Sumner determined a better way in 1837 while at sea.[6] He knew that he was off the coast of Ireland, but had been unable to determine his ship's position because clouds had kept him from making observations of the Sun or stars.

Many methods for determining position are *iterative*. You start with an *assumed position*, that is, you assume that you are at a certain point on the globe, and you use your observations to make corrections to the inevitable error such an assumption introduces. Sumner was using such a method when he was able to get a single observation of the Sun through a break in the clouds.

The method in use[7] would have given Sumner his ship's longitude if the latitude were known. Since the latitude was not known, Sumner solved for longitude three times, using three different assumed latitudes. When he plotted the three "positions," he noticed that all three were on a straight line on his Mercator chart. See Figure 3-1.

Sumner realized that the positions on the line he drew represented the possible places where the Sun's altitude was what he observed when he observed it. Turning this around, he realized that his ship must have been somewhere along that line. Since the line also passed through a well-known lighthouse off the coast of Wales, he set his course to follow the line and thus reach a known location.

Sumner's line is what is now called a *line of position*, abbreviated *LOP*.[8] The important thing is to notice is that if you have *two* lines of position, you know that you are at their intersection. Thus, rather than separate latitude and longitude calculations, you instead calculate two lines of position and determine both latitude and longitude at once.

In a more modern context, all pilots are taught to make a *cross fix* from two lines of position.[9] This is done on a chart: the navigator measures the bearing from a charted object (radio facility, mountain, bridge, whatever) and draws a line from the object with the

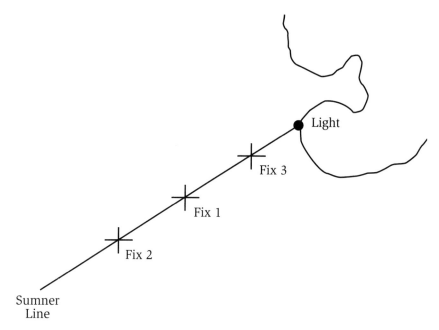

Figure 3-1. Sumner's three "fixes."

proper bearing. Then another line is drawn, based on a different object. Where the two lines meet is ideally the "fix," although the general practice is to use three bearings. This leads to a more reliable fix.[10]

Bowditch says that "The modern navigator thinks primarily in terms of lines of position, rather than latitude and longitude."[11] All of the celestial techniques we discuss will be techniques for determining an *LOP*.

Aspects of Sumner's Discovery

There are two other notable aspects of Sumner's discovery. The first is that a chronometer (a very accurate clock) was needed in order to determine the line of position. Some authors have dramatized the development of the chronometer as a race between the mechanical and the mathematical.[12] But Sumner needed both, and, as you will see, the accurate measurement of time is fundamental for GPS and other satellite systems.

Second, this discovery was made by a man of action, at sea, and under pressure, rather than by a mathematical theorist sitting in a warm study and contemplating. Ordinarily this kind of idea comes from a boffin[13] and wins acceptance slowly, if at all. Oddly enough, although Sumner was obviously not a back-room scientist, his technique and its refinements were not always received with enthusiasm. One well-respected contemporary[14] called them "fancy" and "not suited for oilskin weather."[15]

Sumner's method was refined and improved through contemplation, but each refinement had to stand the test of "oilskin weather," and many were found lacking.

Celestial Navigation and Emergency Training

There is another reason to study celestial navigation: it can save you in an emergency. During my rather short flying career (3000 hours over 18 years), I have probably had about the average number of abnormalities, none of them catastrophic (although some might have become so), and some of them pilot-induced. Airplanes are complicated devices, and just about everything on one can fail. That's why we have backup systems whose use makes up a major part of pilot training.

It seems unlikely to me that someone will learn a new technique in an emergency. We practice scenarios like engine and component failures for this very reason. But many pilots I know spend little time preparing for navigational failures. They pull the preprogrammed route out of the GPS box and follow it. They don't identify navaids. They don't tune any extra navaids. They don't plot fixes. They don't estimate how long it will take to reach the next fix or note when the fix was reached. Where will these pilots be when their navaids fail? Hopefully they'll be in radar contact. If not, how will they find a runway?

It seems unlikely that one will master a new or forgotten technique in a crisis, so I humbly offer this advice: navigate all the time, workload permitting. On a middle-of-the-night lifeflight with a cranky autopilot and a patient on board, I certainly don't do any extra navigation; but on an empty leg when the autopilot is doing its job, I plot NDB bearings, note the stars, and keep a navigation log. I like to think that when I lose everything in the clouds some moonless night I will still be able to get home.

Celestial Coordinates and Time

The stars and planets appear to move on a sphere of undetermined radius around the Earth, which is called the *celestial sphere*. In order to keep track of and predict the positions of the stars and planets, we need to have a set of coordinates for this sphere. The center of the celestial sphere is the center of the Earth. This may make it seem like we are assuming that the Earth is the center of the universe, but what we are really doing has nothing to do with the universe itself. We are choosing an origin of coordinates, and, as we have seen above, we can make any point into the origin and adjust our theories for the origin of choice. Of course some origins are easier to work with than others, that is, some require less adjustment, and this is the case for us.

Because the two spheres are concentric, we can choose coordinate axes for each through this common origin. For the Earth, the z axis goes through the North and South Poles (and the center of the Earth, too), and we'll just extend this to be the celestial z axis. The x axis is harder to define—shall we invite Martians and Venutians to attend a conference on the Celestial Meridian?—but for now note that the equator of the Earth is perpendicular to the z axis, as is the equator of the celestial sphere, so we know that the celestial x and y axes are in the plane of this equator. The result is in Figure 3-2.

The radius of the celestial sphere is irrelevant, so spherical coordinates are natural. We should probably debate the relative merits of the ϕ–θ measurements that mathematicians use and the latitude–longitude approach of navigators, but navigational tradition is that the celestial equivalent of latitude is measured north or south from the celestial equator, although the celestial equivalent of longitude is often measured in one direction from the prime meridian—both for good reason.

The celestial equivalent of latitude is called *declination* (henceforth abbreviated *d*), and it is measured north or south from the celestial

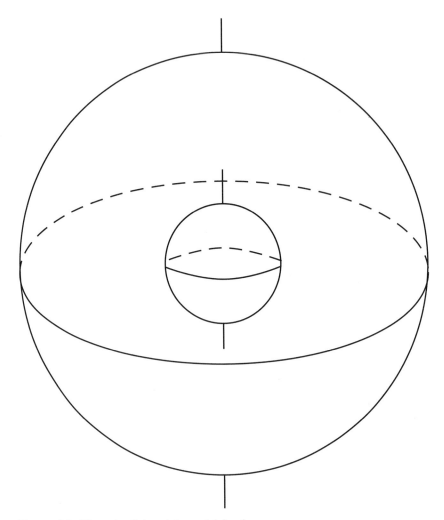

Figure 3-2. The celestial and terrestrial spheres.

equator. Polaris (the North Star) has a declination of approximately 90°, although in fact it varies. The Sun's declination varies between approximately 23° North (at the June solstice) and 23° South (at the December solstice), and it is zero at each equinox.

The celestial equivalent of longitude is called the *sidereal hour angle* (*SHA*). It is measured in a westerly direction. The problem is, westerly direction starting from where? If you imagine the Earth orbiting around the Sun (Figure 3-3), it is just as hard to pick a "celes-

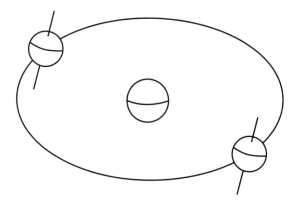

Figure 3-3. Earth in its orbit.

tial prime meridian" as it is to choose an Earth-bound one, and again an arbitrary choice must be made.

But now the arbitrary choice is a point on the sky. This point is called the *first point of Aries*, in reference to the constellation, but this point is in fact moving across the celestial sphere and a new designation must be made every few years. The choice of first point of Aries defines the astronomical *epoch*. The current epoch is 2000.0.

Hour Angles

A celestial object—spaceship, satellite, planet, or asteroid—has a *geographic position* (GP), which is the point at which the line between the object and the origin hits the Earth's surface. The well at Syene was the GP of the Sun, which is why there was no shadow. Since GP is measured relative to the Earth's surface, we measure its θ relative to the meridian of Greenwich, and call it *Greenwich hour angle*, or GHA. In other words, an object's declination is the latitude of its GP, and its latitude is the GHA of its GP.

Usually, almanacs (such as *The Nautical Almanac*) tabulate GHA of the Sun, the Moon, the major planets, and Aries. Stars are tabulated by SHA. To find a star's GHA, you add the SHA to GHA Aries.

It's possible to calculate GHA Aries with fairly good precision. To do this, you need to know the length of a year in days. There are two cases: for a leap year, Aries moves $0.74688984°$, while for a regular year, Aries moves $-0.23875704°$. These are the differences in position on January 1 of each year. On January 1, 1999, at 1200Z, Aries was at $100.2033°$.

The daily movement of Aries comes from calculating at $0.2506847°$ per minute of time; the daily change is $0.2506847° \times 24 \times 60$. (These estimates lead to errors of about $0.2°$ over the course of a year.)

Why are *SHA* and *GHA* called *hour* angles? The reason is that longitude is in many ways equivalent to time. The Earth makes a complete revolution of $360°$ in roughly 24 hours.[16] That means that in one hour, the Earth rotates $\frac{360°}{24} = 15°$. Put differently, if you stand at one point on the Earth's surface, your *SHA* changes $15°$ each hour.

This was the motivation behind the work of Harrison and others to develop the marine chronometer.[17] Let's say that you observe the Sun on your meridian, which happens at local apparent noon. If you also know the current time at Greenwich, then you in fact know your current longitude.

Here's an example. Suppose that your local apparent noon is, according to your chronometer, 7 P.M. Greenwich time. You then make a *time diagram* (Figure 3-4) to help you visualize where you are. This is a circle representing the equator of the Earth as seen from a vantage point far above a pole. Some people use the South Pole, and others use the North Pole; there is no problem as long as east and west on the diagram are interpreted accordingly. Label some point on the circle *G*, for Greenwich.

When it is 7 P.M., it is 19 hours after midnight (navigators reckon time by a 24 hour clock, and refer to that instant as 1900 hours), and 19×15 is 285. This means that, based on our Sun observation, we are $285°$ later than Greenwich. Since the Sun rises in the east, this means that we are $285°$ West of Greenwich. In other words our *GHA* is $285°$ or 19 hours.

This isn't exactly longitude, since we measure longitudes only up to $180°$. To convert a *GHA* of more than $180°$ to longitude, subtract it from 360. Here, $360 - 285$ is $75°$, so based on the observation we decide that our longitude is $75°$ East.

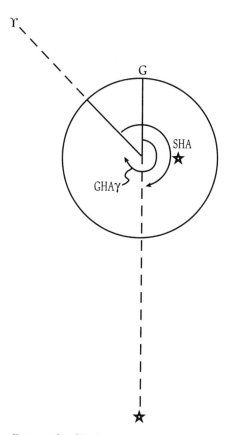

Figure 3-4. Time diagram for Rigel.

Making Chronometers

The difficulty of making a marine chronometer was an engineering problem, not a theoretical problem. For many years, people had understood the principle of the pendulum and other sources of constant rhythm.[18] Ordinary clocks used springs or weights for power, and the pitching of a ship would disturb the weights, while temperature changes affect a spring. Harrison's clock design used a compensating device to eliminate the temperature effects.

Marine chronometers, at least until the quartz revolution, were the subject of almost ritualistic devotion. The chronometer needed to be kept at a point near the center of motion of the ship in order to minimize the effects of movement, and wound the same number of strokes at the same time of each day. Chronometers were not reset, except at intervals when they were cleaned and *rated*, that is, the difference between the chronometer's displayed time and the actual time was measured by an astronomical observatory. A good chronometer does not keep perfect time, since no mechanical device can; rather, a good chronometer has a constant rate. If a navigator knows that the chronometer gains one second every day, then it is an easy enough process to convert the chronometer reading to an accurate time. If, on the other hand, the rate varies from day to day, this conversion is impossible.[19]

Like every other technological discovery, the chronometer was not universally acknowledged, and there was much controversy about its development. (Sobel) outlines some of this history. Joshua Slocum, one of the world's first solo circumnavigators, wasn't quite sure that he saw the value of having his old chronometer rated for his journey, especially since it would cost fifteen dollars to do so!

Time Zones and Q-Codes

The use of time zones was one of the suggestions of the meridian conference, although the rise of railroads in North America had led to some earlier uses. Before time zones, each city had its own time, based on local noon. What would happen to the person submitting a legal document with a deadline of, say, noon at the state capital? In fact, tables of time differences were published. One is on display at the Morristown, New Jersey, Museum.

Before the widespread use of radiotelephone, when all messages had to be sent in Morse code, a scheme was developed to give each of the world's time zones a one-letter name. This is largely forgotten, with one important exception: many people refer to Greenwich time as *Zulu time*. This is because the Greenwich zone was given the letter *Z*, which is spoken as "Zulu" in the international phonetic alphabet. The letters were assigned east from Greenwich, so the next zone to the east is A (or alpha), and so on. The choice of east is appropriate because zone A sees the Sun one hour before Greenwich, zone B sees it two hours before, and so on.

Since the English alphabet has 26 letters and there are 24 zones, who got left out? Only J is missing. M and Y are half zones: M is the western part of the zone split by the international date line, and Y is the eastern half.[20]

There were many short codes like the time zones, several of which are still in use today. Aircraft altimeters measure air pressure and give height above sea level,[21] and they must be reset whenever the pressure changes. Usually one uses the altimeter setting from a nearby airport; this is called *QNH*. But helicopters, gliders, and parachutists like to set the altimeter so that it reads zero on the ground: this is called *QFE*. High-altitude operations are all conducted at standard pressure (29.92˝ of mercury), or *QNE*.

Other Q-codes for aircraft have gone out of use. Many of these were used in direction finding, where an operator on the ground determines the direction from which an aircraft's transmissions come. These include QTE (true bearing of the aircraft), QUJ (no-wind true heading to steer to the station), QDR (magnetic bearing of aircraft), QDM (no-wind magnetic heading to steer), QTF (position of aircraft), and QGE (distance from aircraft to station). QTE differs from QUJ by $180°$; the same applies to the magnetic equivalents.

What Do the Stars Tell Us?

From the perspective of a North American, people in Australia really are upside down. I discovered this on my first night on Australian soil. My head reeled when I first looked up at the night sky: familiar constellations appeared upside down!

Conversely, looking at the stars tells you something about your position. This section will look at the theory of star sights and show why a star sight determines a line of position;[22] the next will look more closely at practice.

Since one nautical mile of latitude is approximately one minute of arc, almanacs and navigators tend to express angles using degrees and minutes, rather than degrees and tenths. Thus, the angle of $41.26°$ is usually expressed as $41°15.6'$. To convert from "dd.ff" for-

mat, multiply the part after the decimal by 60. To convert from "dd°
mm'" to decimal, divide mm by 60.

Some people prefer the even more archaic use of degrees, minutes,
and seconds. In this notation, 41.26° is 41°15′36″ because 15.6′ is
15′36″. To get from decimal minutes to minutes and seconds, multiply the part after the decimal by 60. Many scientific calculators have
an automatic conversion from degrees to this format.

Almanacs and Geographic Position

At any given time, a celestial object has a geographical position (GP)
that we can determine by looking in an almanac. Most of the
almanac is taken up with data about the Sun, Moon, and planets,
because their apparent motion across the sky is so fast. The *Air
Almanac*, published jointly by the U.S. Naval Observatory and the
British Royal Stationery Office, tabulates the 57 so-called navigational stars on the inside front cover,[23] with a more exacting table near
the back of the book. (The *Air Almanac* gives the sidereal hour angle
(*SHA*) and declination of the stars, and the navigator must determine
the *GHA* using a time diagram. *GHA* of Aries is given in 10-minute
intervals. The *Nautical Almanac* gives Aries each hour.)

Pick a star; for concreteness, we'll use Rigel, one of the big stars in
Orion, at noon UTC on December 25, 1996.[24] At that time, *GHA* Aries
is 274°16.7′; the SHA of Rigel is 281°23.8′ and its declination is 8°12.5′
South. This means that *GHA* Rigel is 195°40.5′, so its GP is at
164°19.5′ East. This is about 75 nautical miles southwest of Christmas
Island. The time diagram is Figure 3-4.

If we were at the GP, we could look straight up and see Rigel; in
other words, Rigel would be at the *zenith* (Figure 3-5). As we move
away from that imaginary spot in the ocean, the position of Rigel in
the sky gets lower and lower. The measurement we want to make is
the angle between us (*U*), Rigel (*R*), and the center of the Earth (*C*).
This is called the *zenith distance*, even though it is obviously an
angle. But remember, on a circle (or sphere) angular measure can be
determined from a distance along the circle, or vice versa.

The set of places where the zenith distance is the same[25] forms a
circle on the Earth's surface. The radius of this circle (where distance

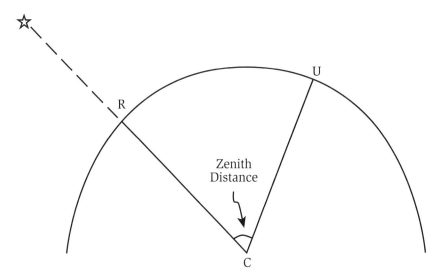

Figure 3-5. Zenith distance.

is measured along the surface of the Earth) is the radius of the Earth
times the zenith distance (expressed in radians).

To illustrate this, let's look at something familiar that can be
expressed as a zenith distance. On a globe, you can see the Arctic
Circle, which is at approximately 63° North. If Polaris were directly
over the North Pole,[26] then at any point on the Arctic Circle Polaris
would have a zenith distance of approximately $90° - 63° = 27°$. So,
the Arctic Circle can be determined as a circle of position based on a
zenith distance measurement.

If you could make two observations, you would know that you
were at the intersection of two such circles. This wouldn't determine
your position exactly, since two such circles generally meet in two
points. But if you had some idea where you were (an assumed position), then you could decide between the two solutions. Or, you
could use a third observation to determine a third circle, which
(except in cases of extreme bad luck or ineptitude) would resolve the
ambiguity. See Figure 3-6.

There are two practical problems with using zenith distances to
determine your position. The first is that zenith distance is difficult

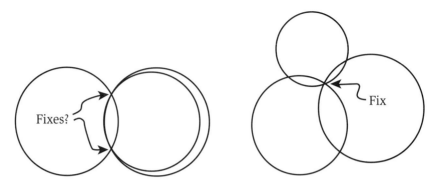

Figure 3-6. Unambiguous and ambiguous three-circle fixes.

to measure accurately because it can be hard to determine which way is straight up. The other problem is that it is difficult to work with circles on the surface of the sphere. If you use a Mercator chart, for example, a circle on the Earth's surface becomes a distorted oval on the chart.

If the zenith distance is large (which it would be for a body near the horizon), then the circle is large, and to draw it you need either a large chart or small scale. We figured out how large the chart would have to be in Chapter 1: more than 80 feet in diameter in order to get the scale currently used in aerial navigation. So a more practical method would use an easier-to-measure angle, lines instead of circles, and would do something about the ambiguity in the position determined.

What Do the Stars Tell Us, Approximately?

One of the keys to using mathematics to solve practical problems is to use approximations, and we have done this many times.[27] A more practical method for interpreting a star sight is to draw the tangent line to the circle of position as a line of position. By definition, the tangent line to a circle at a point P stays very close to the circle if you stay close to P. So, we hope that by using the tangent line we can stay close enough to get a good fix. In practice, the fix we get using the tangent line approximation and a reasonably sized chart is very good indeed.

The zenith distance of a celestial object is the complement of its *altitude*, that is, the angular distance between the object and the horizon. This means, for example, that if an object's zenith distance is $10°$, then its altitude is $80°$. A bubble sextant, which was mentioned in Chapter 1, really measures the zenith distance—the bubble lies along the line from the center of the Earth to the zenith—but the scale is installed in a way that makes it read altitude.

So why don't we use bubble sextants and zenith distances? One reason is that a bubble sextant is less accurate than a traditional sextant at measuring angles, because it is more easily affected by the motion of a ship or aircraft. I think the other reason is the same historical reason that made everyone think it was necessary to find a explicit method for determining longitude: in the Northern hemisphere, your latitude is the (approximate) altitude of Polaris. So, if you have to choose between building an altitude instrument and a zenith distance instrument, build the one that directly measures what you want.[28]

No matter what kind of instrument we use, the measurement we get is a celestial object's altitude. This solves the problem of the difficulty of measuring zenith distance, although it introduces other difficulties when the horizon is obscured.[29]

A sextant is a delicate instrument, and one can spend thousands of dollars to buy one. I personally use a plastic sextant that cost less than $150. It works well for me, at least standing on a beach and sighting against the sea horizon, but I have never tried to use it on a boat. It's worthwhile to spend some time making practice sights; your first few will probably be grossly incorrect, but with practice they should get better. The action of taking a sextant sight is as important to the navigator as the contemplation involved in reducing it. For a complete discussion of the physical use of sextants, and corrections to sextant observations, see (Bowditch) or (Bauer). The latter also discusses how to buy a sextant.

The Method of Marcq Saint-Hilaire and the Dot Product

We're still faced with the problem of drawing really big circles on charts. The method of Marcq Saint-Hilaire[30] addresses this problem.

His idea was as follows:

1. Pick an *assumed position* (*AP*).

2. *Compute* the body's altitude (H_c) at the *AP*.

3. *Measure* (observe) the body's altitude (H_o).

4. The difference $H_o - H_c$ determines how far the *LOP* is from the *AP*.

This is based on the idea that the *LOP* is perpendicular to the direction from the *AP* to the body's *GP*. If H_o is bigger than H_c, the *LOP* is *closer* to the body's *GP*; a memory aid for this rule is **HoMoreTo**ward; see Figure 3-7.

We know how to determine H_o and the body's *GP*; the missing piece of data is H_c, the computed altitude. We'll do this using the dot

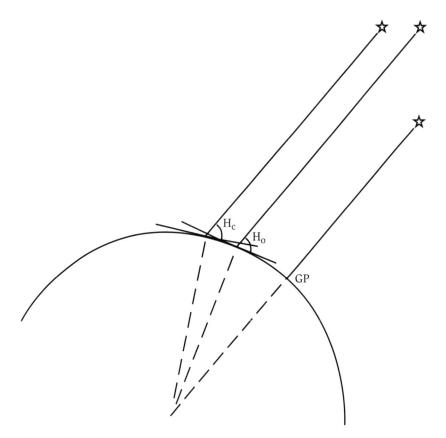

Figure 3-7. HoMoreToward.

product.[31] This makes perfect sense: the computed altitude H_c is an angle, and the best way we have of computing angles is with the dot product.

The angle we want to compute is *GP–O–AP*, where O, as usual, means the center of the Earth. We know the length of the vector from O to *GP* is the radius of the Earth; call this r. The same is true of the length of the vector from O to *AP*. However, the angle between these vectors is the *complement* of the altitude, that is $90° - H$. For any angle H, it is true that $\cos(90° - H) = \sin H$; thus, the dot product of these two vectors is $r^2 \sin(H_c)$. See Figure 3-8.

Both *GP* and *AP* are expressed in latitude-longitude coordinates. We need to convert these coordinates to Cartesian coordinates, like we did in Chapter 2. Let L_a and λ_a be the latitude and longitude of the *AP*, and let L_g and λ_g be the latitude and longitude of the *GP*. Let

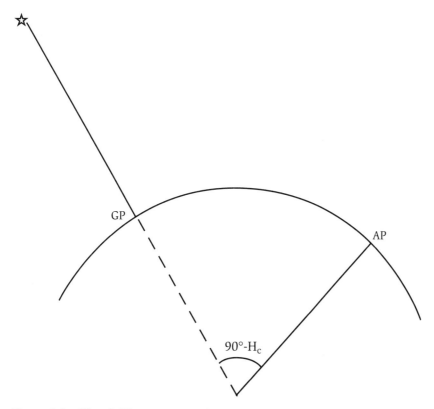

Figure 3-8. *GP* and *AP*.

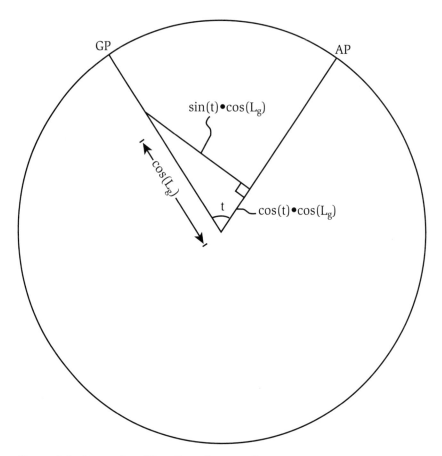

Figure 3-9. Converting *AP* to Cartesian coordinates.

t denote the smaller difference between the longitudes; this is called the *meridian angle*. The *x* axis of our coordinate system will go through the meridian of the *AP*. (Because we will be taking the cosine of *t*, its sign is irrelevant.)

The Cartesian coordinates of the *AP* are (Figure 3-9)

$$(r \cos L_a, 0, r \sin L_a),$$

while the Cartesian coordinates of the *GP* are (Figure 3-10)

$$(r \cos L_g \cos t, r \cos L_g \sin t, r \sin L_g).$$

Taking the dot product, we get

$$r^2 \cos L_a \cos L_g \cos t + r^2 \sin L_a \sin L_g = r^2 \sin H_c\,.$$

The r^2 terms cancel, leaving

$$\sin H_c = \cos L_a \cos L_g \cos t + \sin L_a \sin L_g\,.$$

This is sometimes known as the *cosine formula*.

Now we know the distance from the *AP* to the *GP*, so in principle we could draw a circle of position with that radius. But we have already decided that the circle will be too difficult to draw; instead we want to draw its tangent line.

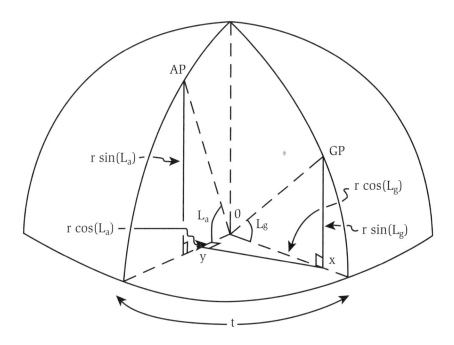

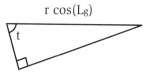

Figure 3-10. Converting *GP* to Cartesian coordinates.

Recall that the tangent to the circle at a point P is perpendicular to the line through P and the center. In our case, this line is the direction of the (great-circle) route from the AP to the GP. This direction is called the *azimuth*. By computing the azimuth we avoid having to draw the whole circle.

Calculating the Azimuth

The azimuth calculation is very important, but it is a little more complicated than the calculation of H_c. Originally, this calculation was done by hand with the aid of logarithm tables. Beginning in 1936, the U.S. Navy began to publish tables of computed azimuth and altitude as H.O. Publication 214, *Tables of Computed Altitude and Azimuth*. Computers were rare then; in fact, "computer" was the job title of the people working on producing such tables. Now, electronic computers are ubiquitous, but many navigators refer to H.O. 214 (or its modernization, H.O. 229, which is still in print) quite often.

So how do we calculate azimuth? First, we need a letter for it: the traditional choice is Z. The angle Z is determined by two planes: the first through AP, O, and the North Pole, and the second through AP, GP, and O. See Figure 3-11.

In Chapter 2, we saw that every plane has a normal direction. The angle between two planes is the angle between their normals, and we can measure this angle by taking the dot product of the normals. In turn, a normal vector can be determined by taking the cross product of two vectors that span the plane.

In the first case, the vectors determined by the North Pole and the AP span the plane; in Cartesian coordinates, these are $(0, 0, r)$ and $(r \cos L_a, 0, r \sin L_a)$, using the same notation as above. But now, since we are interested only in the angles, we will replace these with unit vectors, which will simplify the calculation without changing the result. So we need to take the cross product of $(0, 0, 1)$ with $(\cos L_a, 0, \sin L_a)$; the cross product of these two is

$$\mathbf{v}_1 = (0, \cos L_a, 0).$$

The other plane is spanned by the vectors from 0 through AP and GP; their cross product is a mess. Remember that to find the cross

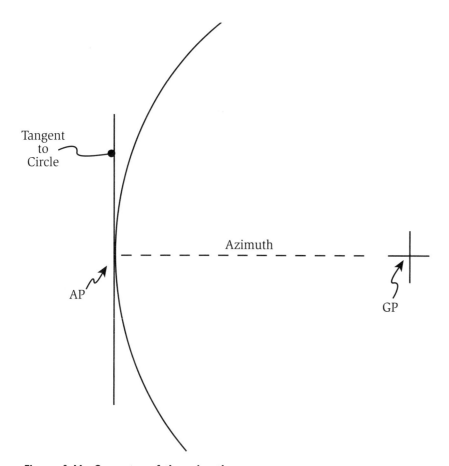

Figure 3-11. Geometry of the azimuth.

product $(x_1, y_1, z_1) \times (x_2, y_2, z_2)$, we can use the expression

$$(x_1, y_1, z_1) \times (x_2, y_2, z_2) = \det \begin{pmatrix} \mathbf{i} & \mathbf{j} & \mathbf{k} \\ x_1 & y_1 & z_1 \\ x_2 & y_2 & z_2 \end{pmatrix}.$$

The two vectors are $(r \cos L_a,\ 0,\ r \sin L_a)$ and $(r \cos L_g \cos t,\ r \cos L_g \sin t,\ r \sin L_g)$; two parallel unit vectors are $(\cos L_a,\ 0,\ \sin L_a)$ and $(\cos L_g \cos t,\ \cos L_g \sin t,\ \sin L_g)$. Their cross product is

$$\mathbf{v}_2 = \det \begin{pmatrix} \mathbf{i} & \mathbf{j} & \mathbf{k} \\ \cos L_g \cos t & \cos L_g \sin t & \sin L_g \\ \cos L_a & 0 & \sin L_a \end{pmatrix}.$$

Mathematicians try to avoid messes by stopping to think before jumping into the calculation. What are we going to do with the vector $\mathbf{v}_2$? We will take the its dot product with $(0, \cos L_a, 0)$, so we will need only the middle ($\mathbf{j}$) component. We conclude that

$$\mathbf{v}_1 \cdot \mathbf{v}_2 = \cos L_a \, (\cos L_a \sin L_a - \sin L_a \cos L_g \cos t).$$

On the other hand, $\mathbf{v}_1 \cdot \mathbf{v}_2 = |\mathbf{v}_1||\mathbf{v}_2| \cos Z$; since $|\mathbf{v}_1| = \cos L_a$, the corresponding factor in the messy form of $\mathbf{v}_1 \cdot \mathbf{v}_2$, is canceled. Furthermore, we know that the length of $\mathbf{v}_2$ is the sine of the angle between the two vectors whose cross product we took. This angle is the complement of H_c, and $\sin(90 - H_c) = \cos H_c$. See Figure 3-12.

Thus, we have determined that

$$\cos Z = (\cos L_a \sin L_g - \sin L_a \cos L_g \cos t)/ \cos H_o.$$

It is assumed that the difference between H_o and H_c is small, so either can be used.

Resolving Azimuth Ambiguity

There is an ambiguity here, since $\cos Z = \cos(-Z)$. This is resolved by the "sign" of the meridian angle t. That is, if the body is west of the *AP*, then the azimuth is measured west; if the body is east, then the azimuth is measured east. If the body is north of the *AP*, then these directions are measured from north; otherwise they are measured from south. See Figure 3-13.

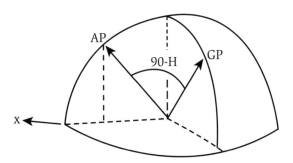

Figure 3-12. Length of $\mathbf{v}_2$.

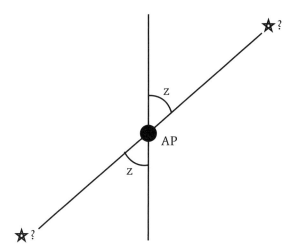

Figure 3-13. Sign of the azimuth.

The Nautical Almanac offers the following procedure for determining the azimuth measured east from north from Z: if $t > 180°$, use Z as is; otherwise negate it.

The Plot

Two such *LOPs* determine a fix at their intersection. This is fine if the observations are taken simultaneously (or nearly so), but if one is delayed the *LOP* from the first must be moved along the chart according to the dead reckoning. In other words, move the assumed position according to the estimated course and speed made good and graphically pull the *LOP* along with it. This is called *advancing the LOP*. If only one body is available, an earlier *LOP* can be advanced and a fix obtained with a second observation.

Usually, three observations are made in practice. This reduces the chance of error. The three bodies should be spread around the compass, because if two azimuths are similar, a small error in one plot moves a long way along the other, nearly parallel, *LOP* and thus leads to a greater error in the fix.

A Complete Calculation

Here is an example of a complete calculation. In practice, a sheet can be made up that directs the navigator through the calculations step-by-step, so that nothing is left out.

A ship off the East Coast of the United States uses an *AP* of 41° North and 70° West (it is common to use whole numbers near the dead reckoning position, originally because the tables are arranged that way and now because it is easier to enter the numbers into a calculator or computer). Two observations were made at 1100 eastern standard time on March 16, 1986: the Sun's altitude of 46°00′ and the Moon's of 24°30′.

First, convert the time to Zulu time, because that's what's in the almanacs. Eastern standard time is 5 hours behind Greenwich, so the Zulu time is 1600 on March 16, 1986.[32] From the almanac, the Sun's declination was 1°40.9′ South and its *GHA* was 57°49.6′. For the Moon, the declination was 23°44′ North and the *GHA* was 352°48′.

By eyeball, the Sun's *GP* is a little to the east of the *AP* and to the south.[33] It's important to keep this in mind as a reality check on our calculation.

More exactly, the meridian angle *t* is 70° − 57°49.6′ or 12°10.4′. Plugging this into the cosine formula (and rounding to four places), we find that

$$\sin H_c = \cos L_a \cos L_g \cos t + \sin L_a \sin L_g$$
$$= (.7547)(.9996)(.9775) + (.6561)(-0.0294)$$
$$= .7374 - .0192$$
$$= .7182$$

so, for the Sun, $H_c = 45°54′$. Since this is 6′ less than the Sun's H_o, the line of position is 6 miles from the *AP* in the direction of the Sun's azimuth.

What is the Sun's azimuth? The formula above will give us the angle east from south, due to the placement of the *GP* relative to the *AP*. Computing to four places as above, we find

$$\cos Z = (\cos L_a \sin L_g - \sin L_a \cos L_g \cos t)/\cos H_o$$
$$= ((.7547)(-0.0294) - (.6561)(.9996)(.9775))/.6959$$
$$= (-.6411 - .0222)/.6959$$
$$= -.6633/.6959$$
$$= -.9531$$

so $Z = 162°23′$; the azimuth is 162°23′ East of North.

For the Moon, $H_c = 24°39'$ and Z is South $79°10'$ East. This yields a fix at North $40°50'$ and West $070°15'$. See Figure 3-14.

Another Graphical Solution

Almost every student pilot buys and is forced to use an *E6-B com-puter.* One side is a circular slide rule set up to do time-speed-distance problems, and the other is set up to do wind triangle problems.[34] This consists of a rotating transparent disk and a grid marked with angles and distances. The grid consists of polar coordinates: the straight lines indicate θ, and the circular arcs indicate r.

Some models of the E6-B come with a square grid and no instructions for its use. The *U.S. Air Force Manual* (USAF) describes how to use this grid for plotting celestial observations and taking a graphical fix. The grommet at the center is the assumed position. For each sight, the azimuth is placed under the "true index" and the intercept is measured up or down from the grommet ("to" is "up").

Once the sights are plotted, corrections are applied (Coriolis and precession-nutation), and the final fix is determined relative to the assumed position. By putting the fix under the "true index," the correction to the assumed position can be determined.

Great-Circle Routes

Given two points on the surface of a sphere, the shortest path between them is a great circle. Many so-called experts[35] naively assert that airplanes fly great-circle routes, but there are many excel-

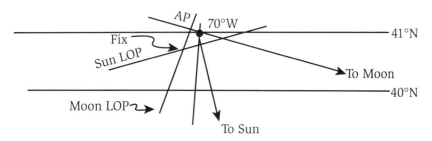

Figure 3-14. Graphical fix from lines of position.

lent reasons for flying a different route.[36] First, a great-circle route is the shortest distance, but it may not take the shortest time because of wind effects. Also, a great-circle route may be impossible to follow for political or geographical reasons. And, a great-circle route involves a continuously changing heading, so it is effectively impossible to follow.

Originally, the use of great-circle routes applied to ocean crossings. Today, aircraft follow a defined system of routes over the oceans. There are a handful of routes between the West Coast of North America and Hawaii; another handful across the North Pacific; and so forth. These routes are fixed. The North Atlantic situation is a little different because the routes change on a daily basis in order to take advantage of the winds aloft. Thus, before flying the North Atlantic, a navigator must find out what routes are in use that day. There is little radar coverage over the oceans, so in order to prevent collisions, aircraft need to follow the assigned track.

Great-circle routes, or at least approximations to them, are now more useful over land, where aircraft are often able to fly so-called random routes. Thus, an airplane flying from Miami to Seattle will typically fly a great-circle route for the portions of the flight away from the airports (where traffic considerations require more control).[37]

There is still the problem of the continuously changing heading needed to follow a strict great-circle route. According to (Bowditch), the standard practice, even at sea, was to select waypoints along the great-circle route and to follow a rhumb line course between the waypoints. A rhumb line is a course with a constant heading.

The Calculation

It's easy, using vectors, to see the great-circle route, especially if the coordinates are chosen with care. Let the meridian angle between the start and end points be t; that is, the difference in longitude between these points is $t°$. Let T denote the start latitude (T for takeoff), and let L denote the end latitude (L for landing).

The great-circle route between these points is in the plane defined by the two points and the center of the Earth. This means that the cross product of the Cartesian coordinates of the points determines the normal to the plane and thus its equation.

To define the course, consider any intermediate longitude m (expressed as a difference from the desired longitude and the longitude of the start). What we need is the latitude M at this point.

The start point has coordinates $(\cos T, 0, \sin T)$, and the end point of the route is $(\cos L \cos t, \cos L \sin t, \sin L)$; the normal to the plane containing the great-circle course is thus

$$\det \begin{pmatrix} \mathbf{i} & \mathbf{j} & \mathbf{k} \\ \cos L \cos t & \cos L \sin t & \sin L \\ \cos T & 0 & \sin T \end{pmatrix}$$

or

$$\mathbf{N} = (\cos L \sin T \sin t, \sin L \cos T - \cos L \cos t \sin T, -\cos L \cos T \sin t).$$

The point along the route at latitude M and hour angle m has coordinates

$$(\cos M \cos m, \cos M \sin m, \sin M).$$

This point is automatically on the sphere, so the only other constraint on M is that the point must be in the plane through the origin normal to $\mathbf{N}$; that is, the dot product of $\mathbf{N}$ and $(\cos M \cos m, \cos M \sin m, \sin M)$ must be zero.

This leads to a fairly messy equation, and, as with all messy equations, some thought is required. First, notice that the components of $\mathbf{N}$ do not depend on m; so rewrite $\mathbf{N} = (N_1, N_2, N_3)$. These can be computed once at the start of the calculation and used for each desired value of m.

This simplifies the equation to

$$N_1 \cos M \cos m + N_2 \cos M \sin m + N_3 \sin M = 0$$

which can be rewritten as

$$\cos M (N_1 \cos m + N_2 \sin m) + N_3 \sin M = 0$$

or

$$\tan M = -\frac{N_1 \cos m + N_2 \sin m}{N_3}.$$

It's instructive to work an example; we'll take the one from Bowditch.[38] The starting point is Manila, at 12°45.2′ North, 124°20.1′ East, and the end point is Los Angeles at 33°48.8′ North, 120°07.1′ West. It's easier to convert to decimal parts of a degree, so the start T is at N 12.753°, E 124.335°, and the end is at N 33.813°, W 120.118°. The meridian angle is $t = 115.547°$. This is easiest to see in the little time diagram in Figure 3-15.

Using these values, $\mathbf{N} = (.1655, .6218, -.7311)$.

Bowditch starts by computing the vertex (see below). This is at W 160°34′, or approximately when $m = 75°$. Since $\sin 75° = .9659$ and $\cos 75° = .2588$, M must satisfy

$$\tan M = \frac{(.1655)(.2588) + (.6218)(.9659)}{.7311}$$

so $M = 41.35° = 41°21′$.

Bowditch then computes latitude at places along the route separated by 12° of longitude (a convenient number for a ship, evidently). These can be arranged in a table, part of which is shown in Table 3-1.

The Vertex

A tradition with great-circle courses is to find the *vertex*, that is, the point closest to the pole. This can be important for ships because of the dangers of floating ice, but it is less important for aircraft (especially in the modern usage, where the whole flight occurs over land). Part of the reason for finding the vertex is *Clairaut's theorem*. Using this theorem makes the vertex the easiest point to calculate and determines an *invariant* of the route: a function that has the same value at every point on the route, much like $x^2 + y^2$ is an invariant for points on a circle centered at the origin.

The vertex is where M is maximal; since the tangent is an increasing function, this is also where $\tan M$ is maximal. Ordinarily, one uses calculus to find the maximum value of a function, but here we

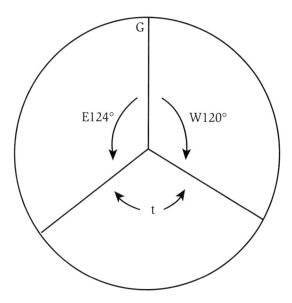

Figure 3-15. Meridian angle between E 124° and W 120°.

can take advantage of the properties of trigonometry functions to find where the maximum occurs without calculus.

The problem is simplified by noticing that the bottom of the expression for $\tan M$ is a constant. So we are left with the problem of finding the maximum value of a function of the form $a \cos m + b \sin m$. The trick is to try to write a and b as trig functions of the same angle; once this is done, we can use the addition formulas to express $a \cos m + b \sin m$ as a single trig function.

Table 3-1 Great-Circle Route

LONGITUDE	m	$\cos m$	$\sin m$	M
Vertex	$\approx 75°$	.2588	.9659	41.35°
W 172.5°	63.1°	.4514	.8923	40.73°
E 171.2°	51.1°	.6280	.7782	38.80°
W 136.5°	99.2°	-0.1599	.9871	38.77°
W 124.5°	111.2°	-0.3616	.9336	35.46°

The values of a and b are not restricted, but in order for them to be the value of sine or cosine they must be between -1 and 1. So, we scale them by a factor of $a^2 + b^2$. This has another effect: the point $(a/(a^2 + b^2), b/(a^2 + b^2))$ lies on the unit circle, so the first is the cosine of some angle and the second is the sine of the same angle. Call this angle p.

We can't just divide through by $a^2 + b^2$, so we have to multiply through by it as well. Thus,

$$a \cos m + b \sin m = (a^2 + b^2)\left(\frac{a}{(a^2 + b^2)} \cos m + \frac{b}{(a^2 + b^2)} \sin m\right)$$
$$= (a^2 + b^2)(\cos p \cos m + \sin p \sin m)$$
$$= (a^2 + b^2) \cos(m - p).$$

The maximum value of cosine (when the angle is between $0°$ and $360°$) occurs at $0°$, so the maximum value of $\cos(m - p)$ occurs when $m = p$.

For our case, $a = -N_1/N_3 = .2264$ and $b = -N_2/N_3 = .8505$. The coefficient $a^2 + b^2$ is equal to .7746, so $\cos p = a/(a^2 + b^2) = .2923$, and $p \approx 73°$.

Approximations

Estimating Azimuth

I have spent some time doing aerial firefighting, which is a confusing and exhilirating kind of flying. On reconnaissance flights, we are often called (in the air) by the dispatch office with a smoke report. Someone calls the local fire department and reports seeing smoke on Sheep Dip Mountain or some other place with an obscure or duplicated name. The dispatch office looks on their maps and determines an approximate latitude and longitude, which they pass on to us.

When this occurs, I know (from GPS) our present latitude and longitude, so I have to figure out which direction to fly to get to the fire. This is exactly the same problem as finding the azimuth: the *AP* is our present position, and the *GP* (which might be better called an *FP*, for *fire position*) is where we need to go. The way I do this is to pretend that the East Idaho Fire District is on a flat Earth instead of a sphere. Latitude is a natural y-coordinate but, because of convergence of meridians, I can't use the longitude as an x-coordinate

Rather than do trig in my head, I use the following approximate method: the difference $L_a - L_g$, expressed in minutes of arc, is the number of miles north or south we need to go: one mile of latitude is one minute of arc. The distance east or west is a little harder. One degree of longitude at latitude L is about $60 \cos L$ nautical miles. Doing a little trig in my head, $\cos L \approx 0.7$, since $L \approx 45°$. This, times the meridian angle (in minutes, of course), is the distance east (or west) we need to go.

Thus, in my part of the world, $\tan Z \approx (L_a - L_g)/0.7t$. For a large range of values, $\tan Z \approx Z/60$.

For example, on patrol at North $42°05'$ and West $111°15'$ (along the east shore of Bear Lake), we receive a call to look for smoke at North $42°30'$, West $112°00'$ (near Sedgewick Peak). Unfortunately, these two places are on opposite sides of the chart.[39] The fire is 25 nautical miles north of our present position and $45 \times 0.7 \approx 30$ miles west. If the bearing is Z degrees West of North, $\tan Z = 30/25$, so Z is a little more than $45°$. A heading of (true) $315°$ will get us pretty close. See Figure 3-16. However, if I continue to fly a heading of $315°$ true, I will travel on a spiral path leading to the North Pole; so, this approximation is good only for short distances.

An Approximate Mercator Chart

The same ideas allow you to make an approximate Mercator chart, which can be useful for plotting celestial sights. This method comes from (USAF), and assumes a spherical Earth. You need a table of *meridional parts* as found in (Bowditch) to construct the chart for other Earth models.

On a Mercator chart, all of the meridians are parallel, so you start by picking a scale s for 1 minute of longitude; that is, let s units be the length of 1 minute of longitude. Draw the whole-degree meridians you want evenly spaced, $60s$ units apart.

The circles of latitude are parallel on the chart too, but their spacing varies. At small scales, each minute of longitude is $\cos L$ times

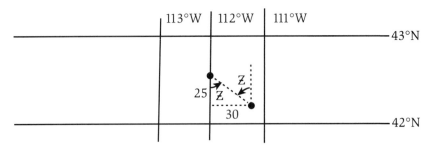

Figure 3-16. Trigonometry and fire patrol.

the length of a minute of latitude. Since we are keeping a constant scale of longitude, each minute of latitude should have length $s/\cos L$. This is OK over small distances, but over a whole degree the variation of $\cos L$ becomes significant, so we will use the average of $\cos L$ and $\cos(L+1)$ as the distance between latitude L and latitude $L+1°$.

Graphically, draw the angle $L+0.5°$ up from the latitude line. Note the distance until the next meridian, and draw the next latitude line at this distance above the previous. See Figure 3-17.

Other Things to Do with a Line of Position

Once we know how to get a line of position using a star sight, we are freed from having to compute latitude and longitude, and thus we can do other things with a line of position. This is based on the more modern point-of-view in mathematics that all coordinate systems are equally valid; the tricky part is relating them to each other. In these cases, we are effectively picking a new coordinate system well suited for the problem at hand. On Earth, a coordinate system is determined by where you put "north." We'll call our new version of north "useful north" so we don't confuse it with "true north" or "magnetic north." Following are some applications.[40]

Speed Line

If you take a sight on a star close to your nose or your tail, you can measure your progress along the chosen course. For an object ahead,

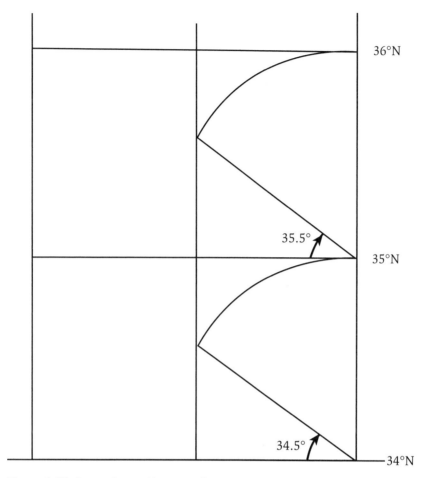

Figure 3-17. Approximate Mercator chart.

you can compare the observed altitude H_o with the computed altitude H_c based on your assumed groundspeed. If H_o is smaller than H_c, then your assumed groundspeed is too high. See Figure 3-18.

This is a very useful kind of sight in an aircraft because it enables you to get an estimate of your groundspeed. This can be critical for long flights over water or harsh terrain: if the groundspeed is reduced, your flight time will be longer, and you are in more danger of running out of fuel.

The effective coordinate system has "useful north" in the direction you are going, and your destination is the effective pole.

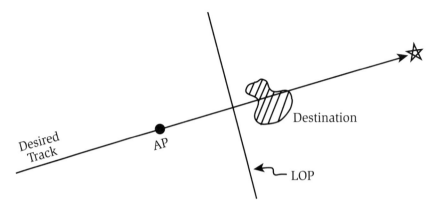

Figure 3-18. *LOP* **from object ahead or astern.**

Course Line

If you take a sight on a star close to one wingtip, you can determine if you are staying on course. Again, comparing H_o with H_c tells you whether you have drifted toward or away from the object, and how far off course you are. See Figure 3-19.

Here, "useful north" is abeam, and you are flying "useful east" or "useful west."

Running Down a Line of Position

If you can determine a *LOP* that passes through your destination, you can "run down the *LOP*" to assure your arrival there. This is what Sumner did, in essence. During a daylight overwater flight, you could compute a Sun *LOP* through the destination, then fly a heading that would intercept the *LOP*. An important strategy in approaching a small target (such as a small island) is to aim slightly to one side or the other; then you know which way to turn along the *LOP* to get there. See Figure 3-20.

The difficulty is that the Sun's *GP* is continuously changing, so its altitude and azimuth are changing. Common practice was to prepare an *altitude-azimuth curve*, which shows the altitude and azimuth of the selected body at the destination for some period before and after the estimated time of arrival. As the destination gets closer, altitudes are taken and compared with the plotted altitude. If the observed alti-

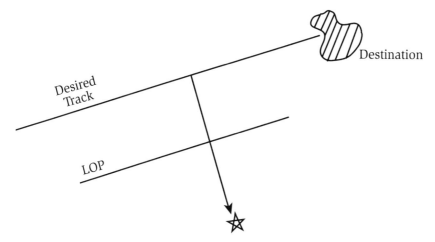

Figure 3-19. *LOP* from object abeam.

tude is too high, the aircraft is off the *LOP* toward the body's *GP*; if the observed altitude is too low, the aircraft is off the *LOP* away from the body's *GP*.

This used to be a difficult task, but today one can write a simple computer program whose input is the observed altitude and the time of observations and whose output is the number of miles left or right of the *LOP*.

What happens during the day? Many days there is only one body visible: the Sun. You can still get usable position estimates by *advancing the line of position*. Shoot the Sun and draw its *LOP*. At a later time, draw a parallel *LOP* at the dead-reckoing position based on the original assumed position. Now shoot the Sun again: the new line will not be parallel to the advanced line and will thus give an estimated fix.

Reducing Azimuth Sights

Celestial navigators, with some exceptions, use the altitude of a body to determine a line of position (*LOP*). Yet pilots (both marine and aeronautical) depend on bearings taken either visually or electronically to determine a *LOP*. The commonplace explanation is that angular measurements along the Earth's surface are, in general, not suffi-

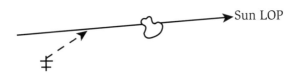

Figure 3-20. Running along an *LOP*.

ciently accurate to be reliable over long distances. How accurate are they? How long is long distance?

Bowditch mentions the possibility of using both altitude and azimuth to determine a one-body fix, although no details are given. This section speculates about calculations a navigator might use to reduce an azimuth observation, and discusses the accuracy of such a fix. The accuracy depends on the altitude H and azimuth Z of the body in an intricate way: low altitudes make the azimuth easier to observe but reduce the accuracy of the fix, while higher altitudes mean a hard-to-observe azimuth with improved precision. Since the determination that high altitudes lead to reduced accuracy is subjective, we can't really quantify the trade-off between the two. And, in fact, high-altitude observations (properly interpreted) are often reduced to highly accurate *LOP*s in practice.

How can a navigator make good use of an azimuth observation? Consider a typical celestial observation using the method of Marcq Saint-Hilaire. The navigator observes the altitude H_o of a body at the time of observation. Using an almanac, he or she computes the altitude H_c that would have been observed at the assumed position (*AP*), and the azimuth Z_c from the assumed position to the body's geographical position (*GP*). The difference between H_o and H_c tells the navigator how far to displace the *LOP* from the assumed position, moving it perpendicular to the azimuth line.

This method takes advantage of the fact that a straight line is approximately a great circle on any small area of a chart made with a Mercator or Lambert projection; the method actually determines a small circle of position.

A straight line also approximates the azimuth line, which is really the great circle between the *AP* and the *GP*. By comparing the observed azimuth Z_o with the computed azimuth Z_c, the navigator could displace the azimuth line parallel to itself and obtain a second approximate *LOP*. The question is how far to displace the line.

The full answer to this question requires the use of calculus, but here are some general considerations. First, if the celestial body is near one of the poles, there will be very little variation in the azimuth with position: after all, Polaris is almost due north from all observers, so no matter how far off course you are, you'll still observe an azimuth somewhere near north.

By contrast, consider the azimuth of a body on the equator. Since $L_g = 0$, the azimuth formula reduces to

$$\cos Z = (\sin L_a \cos t) / \cos H_o.$$

If L_a is zero, the azimuth is 090° or 270°, either by calculation or by analysis, since $\sin L_a = 0$ and $t = H_0$. If instead L_a is 1°, then t is very close to H_o, so $\cos Z \approx \sin L_a$. Since $\sin L_a \approx 0.0174$, Z is approximately $\arccos 0.0174$ or 89°. For small numbers, each degree up or down from the equator leads to one degree of change in the azimuth. Conversely, near the equator, if the observed azimuth Z_o and the computed azimuth Z_1 of an equatorial body differ by d degrees, the azimuth line of position should be moved $60d$ nautical miles.

At higher latitudes, the altitude of an equatorial body is

$$\sin H = \cos L_a \cos t$$

and this can be used to determine $\cos Z$. For example, if $L_a = 30°$ and the meridian angle $t = 60°$, $\sin H = 0.4330$, and $H \approx 25.6°$. Thus $\cos Z$ is about 0.9014 and Z is about 73.9°. Move to latitude 31°, and the altitude becomes 25.4° and the azimuth 73.4°. Thus, each 120 miles of error leads to a one degree azimuth change.

How Well Can One Measure Azimuth?

Now let's address the issue of how accurately azimuth can be measured. The answer is, sadly, not very. The typical magnetic compass is impossible to read to any accuracy better than a degree. On smaller airplanes, the 3.5″ heading indicator may be read to half a degree, with luck. The largest mechanical heading indicators probably aren't much better. The best reading would be with a large *EFIS* compass, which can be programmed to show a short arc; you can read it to an accuracy of $0.5°$. By the examples above, this leaves an uncertainty of about 30 miles or so in your position left or right of the azimuth line.

The error has less effect when H is large. This is exactly the case in which it is most difficult to measure Z_o accurately. However, azimuth bearings are often taken at extremely high altitudes. For example, a mariner taking a bearing off a point at a distance of 6 NM is solving the same triangle as that determined by a sighting of a body at an altitude of $89.9°$. In both cases, it is the geographic position that matters.

Using Azimuth as a Compass

Even when azimuth observations are not accurate enough to determine a fix (for example, the azimuth of Polaris), they are useful as an alternative to the magnetic compass. Several devices have been made for this purpose, notably the sun compass and the astro compass, discussed in (Bowditch). This can be very useful in high latitudes where the magnetic compass is less reliable.[41] Azimuths of this type were the basis of Micronesian navigation; see (Hutchins).

I have sometimes prepared altitude–azimuth curves for night freight flights across the Idaho desert. The azimuth portion is more useful to me than the altitude, since I don't carry a sextant and don't have enough hands to make a sextant observation (night freight flights are generally single-pilot operations, and *someone* has to fly the airplane). An especially nice situation is when a star or planet will be rising or setting over the destination. This gives you information about where you need to steer and some information about your progress along the course: if it rises earlier than expected, you are heading east at a higher speed than predicted.

NOTES

1. This term has occasionally caused some confusion: it means using the stars to navigate, not navigating from star to star.
2. The cockpit voice recording of a recent airline crash showed that the flight crew knew that they were lost yet continued to descend into a mountainous area.
3. Australia has since adopted WGS-84.
4. (Wilford).
5. See (Sobel).
6. (Williams), pp. 112–113.
7. There are many of these, and none of the retellings of this story mentions the exact method Sumner used, perhaps because it isn't relevant.
8. Strictly speaking, the locus of points where the observation might have been made is a circle, not a line; the "line" in question can be thought of as a segment of the tangent line to the circle.
9. At least all pilots in the United States and Canada; some countries do not teach electronic navigation techniques to beginners. British pilots tell me that they don't learn these techniques until they begin to study for an instrument rating.
10. The at-sea practice of this art is described in (Hutchins).
11. Article 2103 of the 1962 edition.
12. For example, (Sobel).
13. For example, the Jimmy Stewart character in the movie *No Highway in the Sky*.
14. See (Williams).
15. One has to wonder what kind of observations might have been made in oilskin weather.
16. It was once believed that this period defined 24 hours and was the most accurate of all possible clocks. Nowadays we know that the Earth's rotation is irregular. See (Landes).
17. See (Landes) and (Sobel).
18. This was how gravitational anomalies were discovered during the French Academy expedition to measure a degree of longitude in Peru; the lighter-than-expected gravity made the pendulum clocks run at the wrong rate (Wilford).
19. See (Bowditch) for more on chronometers.
20. See (Bowditch).
21. See Chapter 5.
22. Actually, it's a circle.
23. See Appendix.

24. Neither Rigel nor any star other than the Sun is visible at noon in England except under very unusual circumstances, so we will also imagine that you are someplace (for example, the Pacific Ocean) where it is night at noon UTC.
25. Mathematicians call this a *locus*.
26. It isn't; the *Air Almanac* contains a separate table for Polaris.
27. A famous industrialist reportedly said, "I don't need it perfect; I need it Thursday."
28. The sometimes-awkward, sometimes-useful measurement of parts of angles in minutes rather than hundredths certainly makes it desirable to avoid subtraction.
29. Every silver lining has a cloud.
30. A French commander; this work is from 1875. See (Bowditch).
31. Traditionally, this is calculated using the navigational triangle and spherical trigonometry. See the Introduction for why we compute the same angle using a different method.
32. A common error here is to use the wrong date; for example, 2200EST on March 16 is 0300UTC on March 17.
33. In an emergency, this may be enough for a rough fix; see the next section.
34. See Chapter 5.
35. These are people of contemplation but no action.
36. See (Portney) for route selection of the *Graf Zeppelin* on its circumnavigation.
37. Airlines are required to use *dispatchers*, part of whose job is to do navigational planning. The flight crew's job is to follow the plan as circumstances allow. If changes are necessary, they are often discussed with the dispatch office. A dispatcher is supposed to consider weather when picking a route, and may send an airliner well off the great-circle route in order to fly in favorable wind conditions. See Chapter 6.
38. 1962 edition, paragraph 822.
39. The standard method for dealing with this situation is to use two copies of the chart. You don't need a complete copy, just the destination and enough of the border area to line things up. This is necessarily an approximation.
40. Another useful coordinate switch, *grid navigation*, is used in the polar regions, although it is not based on *LOPs*.
41. See Chapter 5.

CHAPTER 4

Making Our Own Stars: GPS

Introduction

Many stories about the difficulties of celestial navigation have a common thread: sometimes it's impossible to take sights. Phrases like "due to persistent overcast …" and "due to the lack of a visible horizon …" occur frequently. This is the big drawback of celestial navigation: to use it, you generally need to be able to see both the sky and the horizon in the same direction.[1] Most of us can probably think of a time when we went for days without seeing the sky at all, due to persistent clouds, even when far from the sea. Worse, the sea is an excellent source of moisture, so clouds are plentiful.

The *Global Positioning System*, or *GPS*, is not subject to these problems and therefore provides a reliable and accurate means of position fixing at all times and at all places near the Earth. The reliability is not perfect: engineers refer to "five 9s" reliability, that is 99.999 percent. There are approximately 100,000 crossings of the North Atlantic each year, so it is probable that at least one crossing experiences a GPS problem each year.

The system is expensive (a recent launch cost $80 million, split about evenly between the cost of the rocket and the cost of the satellite[2]), but it works, and most people think the expense is worth it. This section will not go into a lot of detail about the number of satellites, the navigation messages the satellites broadcast, or the like; these are important engineering details but have no effect on the navigator. Instead, it examines the mathematics of GPS. Chapter 5 will discuss its uses.

Even with its incredible accuracy and reliability, there is something a little unsatisfying about GPS. The mathematics is simple, but the engineering is hard. In celestial navigation, a human observer is able to use fairly simple instruments (chronometer and sextant); the same applies to pilotage. But no human being can make the measurements GPS uses. The eye is unlikely to spot the satellites (especially during daylight hours), and the ear cannot hear the ticking of the atomic clocks. GPS seems to be strictly a computer-based endeavor.

How GPS Works

All of our previous measurements have been of angles.[3] GPS measures time, and from this a measurement is inferred.

GPS satellites are in orbits of approximately 25,000 kilometers in radius; they are not geostationary. The satellites are closely tracked by ground stations,[4] so their positions are precisely known. Each satellite carries an atomic clock[5] and broadcasts the time.[6]

The receiver also has a clock, and when a satellite broadcast is received, it compares the time at the satellite to the time at the receiver. From this, the receiver calculates how long it took for the signal to travel from the satellite. Since distance equals rate times time, and the signal ideally travels at the speed of light, the receiver is able to infer the distance to the satellite. Since there are certain errors in this distance, it is called a *pseudorange*.

A pseudorange measurement puts the antenna on the sphere whose center is the satellite's position at the time of broadcast (which the receiver can calculate) and whose radius is the pseudorange (Figure 4-1).

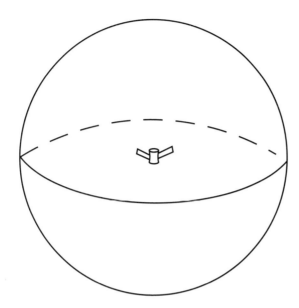

Figure 4-1. Sphere of position.

Two pseudorange measurements place the antenna at the intersection of two such spheres, as in Figure 4-2. This intersection is a circle in space.

Three pseudoranges place the antenna at the intersection of three such spheres; now we are down to two points in space, as in Figure 4-3.

Finally, a fourth pseudorange determines the position of the antenna. This is a position in (x,y,z)-space, centered at the Earth's center. The advantage is that altitude information can be inferred; the disadvantage is that another calculation is required to determine the antenna's geographic position.

Lacking four pseudoranges, a general idea of the receiver's location may also resolve the ambiguity.

Other Features of GPS Receivers

Those who have seen GPS receivers might be a little surprised: they seem to offer so much information. They do, and all of it is inferred from the fix (using the techniques of the previous chapters), which is

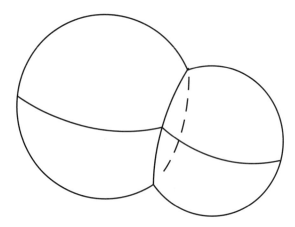

Figure 4-2. Two spheres of position.

inferred from the time it takes a signal to get from the satellite to the receiver. Once you have a fix, you can determine the distance and bearing of any other point using the techniques in Chapter 3.

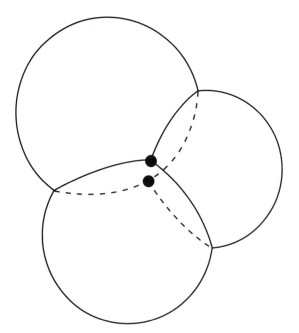

Figure 4-3. Three spheres of position.

Another part of the receiver is its database. Receivers sold for aviation use contain a database of airports, aids to navigation, and the like, and the software to search the database. Advances of this type are based on improvements in computing, not navigation, or they constitute applications of computing to navigation. It is very nice to be able to input "KSLC" to the receiver;[7] the receiver finds the information in the database, including the airport's latitude and longitude, and computes the bearing and distance, which are then displayed.

How GPS Signals Travel

How long does it take for the signal to reach the receiver? This is an application of the law of cosines. Suppose that the altitude of the satellite is H, and look at the triangle whose vertices are the satellite, the receiver, and the center of the Earth. Let ρ denote the pseudorange. See Figure 4-4.

Let R_e be the radius of the Earth (3438 nautical miles) and let R_s be the radius of the orbit (approximately 13,700 nautical miles). At the center of the Earth, one leg of the triangle has length R_e; the other has length R_s. By the law of cosines, then

$$\rho^2 = R_e^2 + R_s^2 - 2R_e R_s \cos L.$$

Table 4-1 shows values of the pseudorange (ρ) in nautical miles as a function of the co-altitude L. This table only goes up to $80°$ since

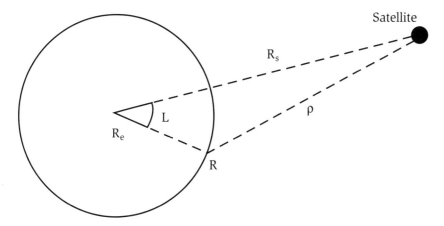

Figure 4-4. Pseudorange and co-altitude.

Table 4-1. Pseudorange and Co-altitude

CO-ALTITUDE	PSEUDORANGE
0°	10249
10°	10318
20°	10522
30°	10847
40°	11272
50°	11775
60°	12333
70°	12922
80°	13521

the satellite signal is not considered useful until it is well above the horizon.

A typical figure from this table is 12,000 nautical miles, which is about 21,900 km. The speed of light is approximately 299,800 kilometers per second; so, the signal takes 21900/299800 ≈ .073 seconds to reach the receiver.

Making It Better

Satellite signals are incredibly accurate, but there is no such thing as perfect accuracy. Engineers have put a lot of effort into improving GPS.

What are the sources of GPS error? The first is on the satellite itself. The atomic clocks are very delicate and must be closely monitored. In fact, several have failed on orbit; that's why there are spares.

Clock failure on orbit is a dangerous situation for the navigator using GPS, because there is bound to be some delay before the ground controllers catch the error and turn off the satellite.[8] During this period your receiver gets an inaccurate pseudorange and you get an inaccurate position.

That's why many receivers (and all IFR receivers) track more than the four satellites strictly needed for a fix. A *receiver autonomous integrity monitoring* or *RAIM* algorithm examines the solutions and determines whether one of the satellites is in error.

One simple RAIM idea needs five satellites. Five positions are calculated, each using four of the five visible satellites. If satellites A, B, C, D, and E are used, the positions are calculated using $ABCD$, $ABCE$, $ABDE$, $ACDE$, and $BCDE$. If these five agree within a certain tolerance, the signal is deemed reliable.

A receiver knows the satellite orbits and is able to predict whether enough satellites will be visible for RAIM at the destination at the predicted time of arrival. For instrument let-downs this is essential: without RAIM, the receiver may unknowingly use bad information and steer an aircraft into a mountain. RAIM problems are rare, but they do occur. I have had RAIM warnings on instrument approaches, although luckily only in good weather.

Another way to catch this kind of error is if the receiver has the ability to read the aircraft altimeter. Then, the receiver can use its Earth model (the Clarke spheroid or, more likely, WGS-84) to compare the calculated altitude with the altimeter altitude. GPS altitude is not very good (50 meter error is common), but a large discrepancy is a sign that something is wrong. This use of barometric altitude is a little like pretending that there is a satellite at the center of the Earth.

Ionospheric Error

Another major source of GPS error is the ionosphere, a layer of charged particles above the atmosphere. These particles change the speed of the GPS signal and, therefore, change the pseudorange calculations. The ground controllers try to measure ionospheric delay, and their estimate becomes part of the message sent by the satellites. These estimates are rather crude, but they are close enough so that there is little effect on aviation.

Differential GPS

The need to correct for ionospheric delay and selective availability[9] led to the concept of *differential GPS*. A differential system has a receiver at a known location and compares the GPS pseudoranges to calculated pseudoranges, in much the same way a navigator doing a celestial observation compares the calculated with the observed altitude. Since the location is known, the differential unit can determine the pseudoranges, and using this information, other nearby receivers can apply these corrections to their pseudoranges. Differential GPS is shockingly accurate, although it is not used much in aviation.

A further refinement is possible with the concept of a *pseudolite*, that is, a pseudo-satellite, placed at a known location. A pseudolite broadcasts a GPS-like signal and provides a very accurate pseudorange. Pseudolites are an integral part of the proposed *wide area augmentation system* or *WAAS* that will provide extremely accurate and reliable[10] vertical and horizontal navigational information for airport approaches.[11]

Intentional Interference

There is another important source of GPS error: military jamming. GPS is a military system, and the military needs to experiment with ways to jam or distort the signals. These experiments are not kept secret, but information on them is kind of obscure. Pilots who check NOTAMs (that is, NOtices To AirMen) can find them under the location "GPS." A typical GPS NOTAM warns that GPS will be unreliable over a certain area (often centered somewhere in Nevada) over a certain time period. Here is a typical example:

> !GPS 09/014 ZLA GPS UNRELIABLE WITHIN THE ARE BOUNDED BY: 3900N/11840W,   3550N/11550W,   3230N/11550W, 3350N/11855W, 3830N/12000W (THIS AREA IS CONTAINED WITHIN A 250NM RADIUS OF CHINA LAKE VOR) AT FL400. THIS AREA DECREASES WITH ALTITUDE . . .

This is a NOTAM issued by Los Angeles Center (ZLA) indicating a large area of GPS unreliability centered at the China Lake Naval Weapons facility. The area is specified by latitude and longitude: 3550N/11550W means N35° 50′, W 115° 50′.

The prudent navigator looks for these warnings and uses another source of navigation to provide an autonomous warning.

Intersecting Spheres in Space

A receiver that knows two pseudoranges can determine the circle in space on which it lies: the circle of intersection of the two spheres. There are many ways to do this. Some ways are easier to understand; some are easier to calculate. The obvious approach to this problem is to take the equations of the two spheres in x, y, and z coordinates and do some elimination. The calculation is difficult to set up in a general way, and the result is hard to interpret. The following method is easier to visualize, although it is not well suited for practical calculations.

The circle of intersection lies on a certain plane in space, which is perpendicular to the line between the two centers, as in Figure 4-5. In order to determine the circle of intersection, we need to find its center and its radius. The radius is the length of the perpendicular dropped from a point of intersection onto the line between the centers. The foot of this perpendicular is the center. Call the radius h and the distance from one of the sphere centers to the circle center x, shown in Figure 4-6.

Let the two radii be r and s, and let the corresponding opposite angles be R and S. Let ρ be the distance between the centers, and let T denote the third angle. Then, by the law of sines (derived in Chapter 5),

$$\frac{\sin S}{s} = \frac{\sin R}{s} = \frac{\sin T}{\rho}.$$

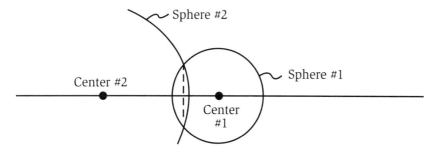

Figure 4-5. Normal to the plane of intersection.

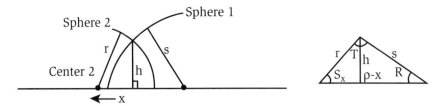

Figure 4-6. Radius of the circle of intersection.

We know all of the bottoms and none of the tops.

There's another equation available: the law of cosines. This tells us that

$$\rho^2 = r^2 + s^2 - 2rs \cos T.$$

The only unknown in this equation is T, which we can solve for. Then we use the law of sines[12] to determine R and S.

Now we know all the angles, so we can determine h as either $s \sin R$ or $r \sin S$; these are equal. The value of x is either $s \cos R$ or $r \cos S$, depending on which sphere center we are measuring from. To be definite, suppose we are measuring from the center of the sphere of radius s; call this point C.

Let **D** be the vector whose direction is from the center of the sphere of radius s to the center of the sphere of radius r, and whose magnitude is x. Then the center of the circle of intersection is just **C + D**.

An Example

Suppose that the spheres are of radius $s = 3$ and $r = 4$, and that the distance between their centers is $\rho = 6$. By the law of cosines, $6^2 = 3^2 + 4^2 - 24 \cos T$, $\cos T = 11/24$, and the angle T is approximately $62.7°$. We don't need this to determine $\sin T$; by looking at the triangle in Figure 4-7, we see that $\sin T = \sqrt{455}/24$.

Since $\sin T = \sqrt{455}/24$, we can apply the law of sines to see that $\sin S = 2 \sin T/3 = \sqrt{455}/36$ and that $\sin R = 3 \sin T/6 = \sqrt{455}/48$. Therefore the radius of the circle of intersection is $s \sin R = \sqrt{455}/12$. The distance from the center of the circle of radius s to the center of the circle is $s \cos R$. Using the same method as above, $\cos R = \sqrt{841}/36 = 29/36$; so $s \cos R = 29/12$. Notice that it was never

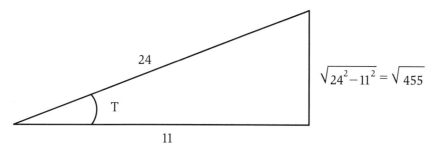

$$\sqrt{24^2 - 11^2} = \sqrt{455}$$

Figure 4-7. Applying Pythagoras.

necessary to evaluate $\sqrt{455}$; we just used the relationships between the trig functions. In the end, we were able to get an exact answer.[13]

Another Approach

We don't need to know the two points where the three circles intersect; we really only need one of them. The fourth pseudorange will determine which of the two we need.

Here's a simpler problem: the three circles intersect in two points, and those two points define a line; let's find an equation for this line. Once an equation is determined, we can intersect the line with the fourth sphere and determine our location in space.[14] Or we could determine another line by selecting a different set of three satellites, and intersect the two of them. Errors may lead to lines that don't intersect, so we would have to approximate the solution by choosing the point where the lines are closest.

We have already determined the plane of intersection of two spheres, because we have its normal (the vector from one center to the other) and (at least) one point on the plane. As we saw in Chapter 1, if the normal vector is **N** and the known point is **P** then vectors **X** on the plane satisfy $\mathbf{N} \cdot (\mathbf{X} - \mathbf{P}) = 0$. If there are three spheres, we have three such planes. In nice cases, we can solve for the line of intersection, but in general special techniques are required because the system is overdetermined.

In the real application, the problems of overdetermined systems and errors in the measurements become more important than the simple geometry of the situation. A human being can't really do these calculations with any speed. GPS requires a computer to do the calculations as well as to make the observations.

Solving for GPS Position on the Earth

As you saw above, GPS gives the position of the antenna in Earth-centered (x, y, z) coordinates. Chapter 1 showed how to convert from these to latitude and longitude. This gives you the *geocentric latitude* L_G. This idea and the associated model of the Earth's shape form the *World Geodetic System of 1984*, or *WGS-84*.

The way GPS derives latitude and longitude is very different from what came before. Before GPS, a *datum* was chosen, and its latitude and longitude were derived. Presumably the latitude—either astronomical or geodetic—was measured very accurately, and the longitude was determined using astronomical techniques.[15]

Surveyors can begin to work after the choice of datum. Survey results depend on the model of the Earth. Suppose a surveyor is at a point exactly one nautical mile due north of the datum. What's the latitude of that point? The model of the Earth as a perfect sphere gives a different latitude than, say, the Clarke spheroid.

The same applies to longitude. The longitude of a point one nautical mile due east of the datum (along a great-circle course) can be calculated if based on the latitude of the point (which is *not* the latitude of the datum, unless the datum is on the equator). So, again, the surveyor's model of the Earth has an effect on the result. Any error in the measurement of the latitude and longitude of the datum is repeated in every latitude and longitude measured with reference to the datum.

But GPS latitude doesn't depend on a model because it is geocentric. Before GPS, aeronautical charts in the United States were based on the Clarke spheroid and the North American Datum of 1927. The datum was at Meades Ranch, Kansas.[16] Each country used its own datum, and there was really no way to compare these until GPS and the practical ability to use the center of the Earth as a reference point. The British Ordnance Survey datum differs from

WGS-84 by about 700 meters; the Japanese datum differs from WGS-84 by almost twice that.[17]

When the United States switched from NAD-27 to WGS-84, many locations "moved." The continental United States saw generally small movements, but some of the changes in Alaska and Hawaii were substantial.

Since Japan uses its own datum, there has been a lot of confusion. It's possible for a navigator to know that the Sendai airport is at N 38° 08′, E 140° 57′ and that the FMS says that the aircraft is at N 38° 08′, E 140° 57′, but for the aircraft not to be over the airport. This is because the airport coordinates are based on the Japanese datum while the aircraft coordinates are based on WGS-84.

After the latitude and longitude are determined, the GPS receiver also needs a model of the Earth in order to determine altitude. The receiver calculates the value of sea level at the desired location, and compares it with the distance from the derived (x, y, z) position to the center of the Earth. The difference is altitude.

An Example of Altitude

In the United States, the National Oceanographic and Atmospheric Administration (NOAA) maintains a list of geographical reference points. This is available on the Internet[18] or in published form.

The Airport Reference Point for the Pocatello, Idaho, Regional Airport is included in this database. The coordinates are given *in terms of NAD-83* (which is equivalent to WGS-84 for this purpose) as North 42° 54′ 47.03813″ and West 112° 35′ 26.36316″. Since one second of arc (1″) is 1/60 of a nautical mile or about 100 feet, .00001″ is about one-one-thousandth of a foot.

The elevation is given as 1357.7 meters or 4454 feet. The vertical reference is the North American Vertical Datum. The ellipsoid height at this location is given as 1344.73 meters.

The database also includes the *x*, *y*, and *z* coordinates with reference to the Earth-centered frame. These are (in meters) $x = -1797612.842$, $y = -4320473.246$, and $z = 4321349.222$. We can use this to see how the Earth's shape differs from a sphere.

First, for a spherical Earth, z should equal the radius times the sine of the latitude. The sine of the latitude (to 6 places) is 0.680888, so the equivalent radius is about 6347 kilometers.

Another way to get the radius of a sphere is by calculating $\sqrt{x^2 + y^2 + z^2}$. In this case, the "radius" (actually, the distance to the center of the Earth) is about 6370 kilometers.

Finally, on a spherical Earth, x should be $r\cos(L)\sin(\lambda)$, where λ is the longitude, and y should be $r\cos(L)\cos(\lambda)$. The value of $\cos(L)\sin(\lambda)$ is about 0.2813, while $\cos(L)\cos(\lambda)$ is about 0.6762. The first would imply that the radius is 1797612.842/.2813 or 6391 kilometers, and the second implies a radius of 4320473.246/0.6762, or 6389 kilometers.

NOTES

1. There are sextant techniques for using a horizon directly behind you, too; or you can use a bubble sextant. But most mariners use optical sextants, which need the horizon as a horizontal reference.
2. *Aviation Week*, July 1997.
3. Or of time, which is equivalent to an angle in the context.
4. This is expensive. Still, the GPS-consuming public thinks of it as a "cheap" system.
5. Or several; the details are unimportant here.
6. Among other things; again, the details are unimportant here.
7. That is, Salt Lake City, UT. The "K" indicates an airport in the United States. There are two sets of airport identifiers in the world. The one you see on your luggage tag is from a list compiled by the International Air Transport Association. Pilots use identifiers compiled by the International Civil Aviation Organization. So, while your bags go to NRT (Tokyo, Narita), your flight crew goes to RJAA (also Tokyo, Narita).
8. The satellite is not powered down, but its signal is marked as unusable. The satellite signal can also be marked as unusable for scheduled maintenance.
9. Selective availability (SA) was a GPS "enhancement" that degraded the military signal for civilian use. When SA was on, a non-moving GPS receiver often showed a groundspeed of 1 or 2 knots; without SA, the groundspeed shows zero.
10. Reliability is perhaps of more interest to a prudent navigator.
11. At this writing, the WAAS system has been having software problems and is not operational.
12. Derived in Chapter 5.
13. But see the comments on measurement errors below.
14. There are two such intersections, but in general only one of them is between the satellite and the Earth.
15. In some cases, not very accurately. See (Wilford).
16. I flew over it once.
17. (Featherstone and Langley).
18. Since the Internet is changing so rapidly, I have not included the URL.

Navigation in Airplanes

Introduction

Navigation is both an art and a science. The previous sections have been more concerned with the science. But in order to master navigation, you must do more than sit on the ground and think about geometry and geodesy. To say that navigation is an art really means that navigation takes judgment and experience.

This chapter addresses the practice of navigation in airplanes. It includes both modern and traditional methods. Modern methods are fantastic, but, as Bowditch wrote, "Until such time as mechanization may become complete and perfect, the prudent navigator will not permit himself to become wholly dependent upon 'black boxes' which may fail at crucial moments … [t]he wise navigator uses all reliable aids available to him, and seeks to understand their uses and limitations." The pilot who lets the flight management system (*FMS*) tune all the navigational radios and derive a position while never looking at a chart or the sky has, in a sense, become a passenger in a black box. Its failure at a crucial moment could be fatal.

Approaching Kemmerer, Wyoming, in a FMS-equipped King Air, we were able to descend underneath the clouds into mountainous terrain. Kemmerer has no instrument approach; the airport must be found visually. The black boxes—worth tens of thousands of dollars—

become useless at this point, and the pilot without a $7 sectional chart is, quite literally, up the creek. But since the chart was already out and folded to the proper panel, we found the airport easily.

The two young naval officers in the introduction might have been more successful if they had used the LORAN in their daily navigational chores, rather than waiting until a crisis. The accumulated experience might have tempered the captain's skepticism. Conversely, older methods that aren't used at every opportunity become rusty and unreliable, especially in an emergency.

This chapter will discuss some of the equipment an aerial navigator uses; for more details, you should see any of the standard pilot training texts or *The Pilot's Handbook of Aeronautical Knowledge* (Illman).

Advantages and Disadvantages of a Database

The navigator has two tasks. The first is to determine the present position, and the second is to determine how to get from here to the destination. One of the nice things about electronics is their ability to store a database of "theres." A pilot who has been cleared to a fix called OBSCR (pronounced "obscure") can type[1] the five-letter name into the black box, and the black box will calculate the distance and bearing.

Judgment and experience show that there are several sources of error here. First, the navigator could blunder and type in OBCSR; the black box would think this was just fine. There might be two fixes in the database called OBSCR; this is less likely with a five-letter name, but the existence of two fixes with the single-letter identifier *R* was a factor in a major airline accident. A fix near my home airport in Idaho has the identifier *PI*, as does a fix in Florida. This is a more systematic error.

This wasn't a problem before computerized databases. If a navigator tunes a low-frequency radio and hears the Morse code identifier *PI*, it is very likely that he or she will know without doubt whether the station is the PI in Florida or the PI in Idaho, since the range of most stations is less than 100 NM.

There can be more problems if the database is out of date. For various reasons, aeronautical beacons are moved or renamed. Recently, the NDB at Meridian, Idaho, was moved. If the black box has the old coordinates for Meridian, the airplane won't go to the beacon; it will go to the shopping mall (or whatever) built at the beacon's former location. An out-of-date database is an error of carelessness.[2]

Even if OBSCR is correctly entered, the navigator may find it impossible to get there directly. There may be mountains higher than the airplane can comfortably climb; the terrain might be unsuitable in other ways; or there may be some kind of airspace restriction. Still, knowing where OBSCR is can be a big help.

Other Modes of Travel

This section is about airplanes, but the same principles apply to boats, hikers, snow machines, spaceships, and the like. The difference between these is error tolerance. For an airplane in cruise flight, a one-mile error is generally insignificant; for a hiker, one mile is a major error. It's probably best to measure error in time rather than distance, but it's hard to plot time directly on a chart.

The hiker and the pilot use the same general methods. Hiking in Yellowstone Park, we lost the trail back to the car. I took a compass bearing off the peak of the Grand Teton, and determined that we were on the same *LOP* as when I had taken a previous practice sight from a known position. A look at the topographical map showed that a northerly heading would re-intercept the trail from any reasonable spot on the *LOP*. It did.[3] Or was it that I had maneuvered off course in an airplane to get around some rain, noticed that I was on a certain electronic *LOP*, and looked at the map to see that a northerly heading would re-intercept the airway?

Instrument and Visual Navigation

There are, broadly speaking, two kinds of flying: VFR and IFR. This chapter concentrates on so-called *VFR* flying. VFR stands for *visual flight rules*. The other kind of flying is *IFR*, or *instrument flight rules*. A pilot must earn an instrument rating in order to fly IFR. This is in addition to, and for some more difficult than, earning a private pilot certificate. The distinction is technical: each set of rules has different methods for keeping airplanes from colliding. Under VFR, the rule is "see and avoid"; under IFR, a controller on the ground uses radar or other methods to prevent collisions. A pilot may fly VFR if certain weather conditions exist; basically, the VFR pilot must have enough visibility to see other airplanes.

One instrument student said that VFR stands for "vacation fun rules," while IFR stands for "incredibly frustrating rules" because of the extra work involved in fitting into the system with all of the other airplanes. I'm not sure I agree. Suppose you are traveling eastbound and there is a layer of clouds with tops at 8900′ MSL. Flight above the clouds is nicer since the air is generally smoother than it is below the clouds, and in the United States it is legal to fly under VFR "on top."[4] Under VFR, an eastbound flight would be required to fly at 11,500′ to stay on top; you need to be at 9900′ or above because of the VFR cloud clearance rules, and the lowest legal eastbound VFR altitude above 9900′ is 11,500′. But flying that high is uncomfortable without oxygen, and piston-powered airplanes without some kind of supercharging are less efficient at that altitude. Under IFR, the same flight can be made on top at the more efficient (and comfortable) altitude of 9000′, because there is no minimum cloud clearance requirement. I prefer smooth flights on vacation; the constant adjustments needed when in turbulence can be incredibly frustrating.

Smooth flights aren't just for vacation. I know an airline captain who used to fly Navy P-3s on long patrols. During these patrols, one or more of the P-3's four engines would be shut down to save fuel. There were all kinds of official suggestions about what altitude was appropriate for this mission, but the first rule was to find an altitude with minimal turbulence so the crew would be comfortable. If the crew was uncomfortable, no useful work would be done.

Many pilots think that VFR and IFR navigation are different. Techniques that are important in one case can still be useful in the other, except that the IFR pilot can't always count on using *pilotage*— that is, flight with visual reference to objects on the ground.[5] But the basic problems are exactly the same, and there are lots of things on the ground for the IFR pilot to refer to. Most of these are radio aids to navigation, which we will discuss in detail. But a little pilotage is possible: weather radar can be used to "see" mountains and lakes.

And there are little local tricks, like the big sign on the ground that turns the clouds red or the column of rising air over the highway, which some pilots use to help pinpoint position.

The Compass

The most basic and reliable navigational instrument is the magnetic compass. It is a great piece of luck that the Earth currently has a strong and steady magnetic field. This has not always been the case. Geologists have determined that the Earth's magnetic field has gone through many reversals, and that such reversals may take a few hundred years. Thus, there have been eras several generations long during which we had no reliable magnetic direction reference. We're much better off than Martians; the magnetic field of Mars is so weak that it was only discovered by the *Pathfinder* spacecraft in 1997.[6]

Well, most of us are lucky. Near the magnetic poles, the magnetic field is more vertical than horizontal, so the compass needle tends to point up or down rather than north. This affects a few people near the South Pole (there are very few people near the South Pole), but the North Magnetic Pole is close to the major air routes between the Americas and Europe. In the polar regions, you can use the Sun as a direction reference.[7] That's why the written examination for the Canadian commercial pilot license has a question about determining the Sun's true azimuth.

The first aviator who needed a compass was Louis Blériot during his daring first aerial crossing of the English Channel in 1909.[8] Evidently he forgot to look at it, because the course he followed was many degrees off his intended course.

Even if Blériot had looked, he would have discovered that a compass is pretty hard to read in the air. The vertical components of the magnetic field cause the compass to jump ahead or lag behind during a turn or acceleration. In the Northern Hemisphere, the compass lags in a turn while heading north; that is, if you turn right from a northerly heading, the compass initially indicates a turn to the west, which is to the left. It leads a turn from south: turning right from a southerly heading, it indicates a larger turn to the west than is actually occurring.[9] On an easterly or westerly heading, the compass shows a northerly turn if you accelerate in a straight line, southerly if you slow down.

This is why most airplanes and many ships carry a gyroscopically stabilized compass. This has to be set to agree with the magnetic compass. In larger airplanes, this is done mechanically, but in small airplanes the pilot must set the heading indicator by hand. A good one will stay in close agreement for an hour; a bad one must be reset every few minutes. Failing to set the heading indicator is a major cause of getting lost. Every airplane, ranging from one-seat aerobatic marvels to 400-seat wide body airliners, carries a magnetic compass in the cockpit.[10]

Magnetic North and True North

In most cases, magnetic north is not true north. Worse, magnetic north may not even be exactly in the direction of the Magnetic North Pole. There are lots of local magnetic deposits that affect the compass. If you stand halfway between the North Pole and the Magnetic North Pole, your compass will point in a generally southerly direction, but not due south.

The difference between what a magnetic compass reads and what a true compass would read is called *variation*. This is a systematic error in direction measurement, but it is charted on every aeronautical chart I have ever seen, so we can correct for it. Variation changes over time; if you look at older charts, you will see that the isogonal lines, or lines of equal variation, have moved at a rate of a few miles a year.

Edmund Halley, who is more famed for computing the orbit of a certain comet, proposed a survey of magnetic variation as a means of estimating longitude.[11] The idea was to measure the variation by comparing the compass bearing of the Sun to its true azimuth, or by noting the magnetic bearing to the Sun at local noon. Then, knowing the ship's latitude, its position would be at the intersection of the parallel of latitude with the isogon. One problem with this is that some of the isogons are rather far apart, making the fix obtained rather vague.

Plotting a Course

When a navigator measures the course from here to there, the measurement is usually in relation to true north, since most charts include the meridians. This measurement is called the *true course*. For example, the true course from Pearson International Airport in Toronto to Greater Buffalo International Airport is about 136°. The variation in this part of the world is about 9° West, which means that the compass points 9° west of true north. The rule of thumb for correcting for variation is to add westerly variation and subtract easterly variation: East is least, West is best, so the magnetic course is about 145°.

In planning such a flight, I would draw a straight line connecting the two airports on the chart. This is a good way to make sure that the direct route doesn't need to cross a mountain, a large body of water, or a military firing range. If it did, which is true for almost every flight in a small plane in Idaho, I would need to plot some turn points or waypoints to design a route around the problem.

Next to each course segment I would write the true course as "TC136" (I might need this later because winds aloft are forecast with respect to true north). I would then write "MC145" for the magnetic course. Finally, I would write "58 NM" because the length of the leg is 58 nautical miles. If there are multiple legs, I would do this for each leg.

Compass Course and Meridian Convergence

It is generally considered good practice to refer to a course or heading using a 3-digit group. One degree east of north should be spoken or written as "zero zero one." This reduces errors of hearing and the like. Air traffic controllers always use, and always expect, a 3-digit grouping for direction.

In measuring a course on a chart where the meridians converge (like a Lambert chart, see below), you should measure it near the middle of proposed course line. This averages out the differences. This won't work on a really long flight, however; you may need to divide the flight into segments and measure the course for each segment separately.[12] The size of the segment would depend on your speed. A segment length between 30 minutes and an hour would probably be about right.

Charles Lindbergh used this method to plan his route across the Atlantic in 1927. He described his planning in his book *The Spirit of Saint Louis*. He first worked graphically, plotting the great-circle route on a *gnomonic projection*; this is a chart designed to show great-circle routes emanating from one particular location, in this case New York. He then selected waypoints and transferred them onto a Mercator chart. To check, he went through the calculation of the great-circle courses for each leg. (Actually, he found such close agreement with the graphical result that he never finished the calculations.)

Today, aircraft use the the North Atlantic Track System to cross the North Atlantic. It changes daily to take advantage of winds aloft. Each track is defined using waypoints at each 10° of longitude. Since the track system is usually centered at around 60° North, one degree of longitude has length cos 60° NM, or about 30 nautical miles, so the checkpoints are roughly 300 nautical miles apart; this is less than an hour in an airliner.

Compass Deviation

There is another source of error in using the compass: the big metal airplane with a big metal engine and lots of black boxes shooting electrons around is bound to have some magnetic effects. This can also be measured, and it is called *deviation*. Every airplane has a deviation card posted near the compass. It is seldom referred to other than by people crossing oceans or remote areas, who refer to it regularly. People navigating over areas rich with visual or electronic position checks usually assume that the difference between their computed headings and what actually works is due to the difference between the forecast and actual wind. A careful pilot might refer to the deviation card when resetting the heading indicator to agree with the compass; but most seem to accept the 1° or 2° error.

How much error is this? Remember that the tangent of a small angle is about equal to the angle *in radians*, so the tangent of 1° is about $\pi/180 \approx 0.0175$. (The actual value differs from $\pi/180$ by about 0.000002.) A more useful approximation is that tan 1° is about 3/180, since π is pretty close to 3.[13] So, an error of 1° over 120 miles puts you 2 miles off course. Depending on the circumstances, this is probably acceptable, assuming that you have some way to update your position at that time. But an error of 5° over the same distance puts you 10 miles off course; this may put you out of the visibility range of your destination.

There are other times when this error is unacceptable; namely, when crossing an ocean or featureless terrain. If your only navigational instrument is the compass, you need to make sure that you read it correctly and hold the computed heading. This is the case in my 1946 Taylorcraft crossing the Idaho desert, and it's the case in an airliner whose electronics have failed over the ocean. This is not to say that you may not have any drift correction: most such flights start with at least a short segment over land (or within range of ground-based electronic aids), and a prudent navigator determines the drift while this information is available and adjusts the computations accordingly. Aircraft used to have a *drift meter*, which was a hole in the belly with an etched glass that the navigator would rotate so that objects on the ground went along the grid. See (USAF).

Charts

Most aeronautical charts are based on the Lambert conformal projection. The Lambert projection is a conical projection with some changes in the way latitudes are distorted, which make it preserve angles. Any angle you measure on a Lambert chart is exactly the same as the corresponding angle on the Earth's surface. On a Lambert chart, a straight line is also very close to representing a great circle.

The cone defining a Lambert chart meets the Earth's surface along two circles of latitude; these are called *standard parallels*. The distortion is zero at a standard parallel, as we saw in Chapter 1. For example, the Salt Lake City sectional aeronautical chart, published by the National Oceanic and Atmospheric Administration,[14] uses the standard parallels 41°20′ and 46°40′; the latitudes on the chart range from 40° to 44°30′. so no point on the chart is more than 2°10′ from a standard parallel.

In contrast, the Denver chart includes latitudes from 36° to 40°, and the standard parallels are 33°20′ and 38°40′, so the maximum distance to a standard parallel is 2°40′.

The Australian Townsville World Aeronautical Chart encompasses latitudes between 16° and 20°, with standard parallels 16°40′ and 19°20′. No point is more than 1°20′ from a standard parallel. This chart contains a topographic base reliability diagram indicating that the whole chart was compiled from reliable sources; other Australian charts have areas where vertical errors may exist, that is, where you might hit a mountain even though you are above all of the depicted terrain.

The most common scale is 1:500,000. In the United States these are called sectional charts. The German version is called a *Luftfahrkarte Aeronautical Chart*; the Japanese version is called a *Koubun ̀KouKu Zu*, or division (sectional) navigation chart. This scale works well in practice at speeds up to about 150 knots. Since this scale is approximately 7 NM to the inch, a 420 knot airplane[15] covers an inch on the chart in one minute. This is fast enough to have to continuously move your finger to keep it over your present position.

The next most common scale is 1:1,000,000. These charts go under various names, but *World Aeronautical Chart* is used in the United States, Canada, and Australia. The U.S. Department of Defense has charted the whole world at this scale as Operational Navigational Charts.[16]

One warning: some IFR charts aren't drawn to scale. They are explicitly labeled "not to scale." You can't use them to plot a course or for anything more than a rough estimate, such as "I'll have to go west." Since these are used for IFR, there will always be some kind of electronic guidance available.

Charting Details

What do the charts show? This varies from publisher to publisher,[17] but they all include a terrain depiction; bodies of water and rivers; some cultural features such as outlines of built-up areas, roads, railroads and the like; and airports. Bodies of water are important since most aeronautical hazards are caused by water. I try to avoid using airports next to water.

Most charts show *navaids*, that is, aids to navigation, although many do not show the frequency on which these broadcast. They also show such obstructions as towers and cables, although in some cases these go uncharted, at least for a while. Charts of my local area from 1993 show no tower on the side of Bonneville Peak; 1995 charts show a 309´ tall tower with the notation UC, which can mean either "unconfirmed" or "under construction"; 1997 charts show the tower.

This is part of the reason most of these charts have an expiration date, and it is unsafe to use old charts for navigation. In the United States, sectional charts become obsolete every 6 months. In many cases, the changes are minor, such as radio frequencies, but events like the eruption of Mount Saint Helens require changes to the terrain depiction as well. Or there might be a change in airspace classification or boundaries, especially in the vicinity of major metropolitan airports.

Pilotage and Dead Reckoning

Pilotage is flight by reference to objects on the ground: "We're over White Pigeon."[18] The nautical equivalent is used near shore. Good pilotage requires a good chart. In taking my 1946 Taylorcraft, which has no electrical system, from Idaho to the West Coast, I used pilotage: Interstate 80 to Lake Tahoe, then California Highway 50. I followed the same truck from Wells to Winnemucca.

Pilotage is usually combined with *dead reckoning* (some people call this *ded reckoning*, where *ded* is short for "deduced"). If you fly, walk, or row in a certain direction at a certain speed for a known amount of time, you can deduce your position. A fix obtained in this way is usually called a *DR* position. Usually, neither the direction nor the speed is really certain because of the vagaries of wind and current, so a prudent navigator looks at a DR fix with some skepticism. One thing that can improve the quality of the DR fix is a *wind triangle*, described below. By combining DR and pilotage—especially by using pilotage with fairly frequent checkpoints—you can "correct" the DR.

Frequent fixes are very important in VFR flight in reduced visibility. The simplest way to do this is to make an X on the chart and note the time whenever you have a good idea where you are. If you get lost or miss a checkpoint, you can use dead reckoning from your last known location to determine where you are.

For example, we were flying from Cordova, Alaska, to Talkeetna, and the visibility was just about 3 miles—legal but difficult VFR, especially so far from home. As we passed each headland or island, an X was made on the chart at the fix and the time was written next to it. The chart shows a nice procession of Xs at (Zulu time) 1721, 1725, 1726, 1731, 1734, and 1737. Frequent fixes are helpful in low-visibility VFR flight. Our chart is Figure 5-1.

Around 1737 we got distracted by the need to make a radio call for an update on the weather ahead. While we were distracted, the heading wandered from the planned 220 degrees to 190 degrees and stayed there for a few minutes.[19] This is bad. Some land appeared out of the mist. We were expecting to see Esther Island, a very large island separated from the mainland by a narrow channel at the entrance to the Wells Passage, but the DR—8 minutes at 120 knots on heading 190—indicated that it might be Perry Island. We each argued our point forcefully, but circling the island showed that the DR was right.

Useful Shortcuts and Approximations

There is room for judgment in applying dead reckoning, as there is in any form of navigation. It is possible to be too precise, which won't change how you travel but does make you work harder. Flight instructors correctly teach students to calculate to the nearest knot, but how do you answer the passenger who asks, "Are we there yet?" Do you pull out a calculator and answer "26.914 minutes" or "about half an hour"?

The way to answer "about half an hour" is to use good estimates. Trainers go about 90 knots, or 1.5 miles per minute. Most 4-seat airplanes go 120 knots, or 2 miles a minute. High-end 4-seaters and most 6-seaters are good for 150 knots, or 2.5 miles a minute; for working in your head, it's usually easier to think of this as 5 miles in

2 minutes. Turboprops go 4 or 5 miles a minute, and jets in cruise go 8 or 9 miles a minute.

This is simple but surprisingly accurate. Fifteen miles in a trainer takes 10 minutes at 1.5 miles per minute. What if there's a strong headwind and the groundspeed is 80 knots? Then it takes a little more than 11 minutes. What if there is a tailwind, and you can make good 100 knots? That 15 miles will take 9 minutes.

I wouldn't use this except as a rough planning guide for crossing an ocean (Let's see, 2700 miles at 9 miles a minute is 300 minutes or 5 hours, and we carry enough fuel for 6 hours), but I do use it a lot in firefighting. The dispatch office always wants an ETA to someplace strange, right now. Depending on the airplane, I divide the distance by 2 or 3; it seems to keep them happy. If we're in a single-engine airplane, we are required to be on the ground 30 minutes after sunset, so again I divide the distance home by 2 or 3, depending on the airplane, to determine how many minutes of flying time I need to get home. I usually add 5 or even 10 minutes as a contingency.

Airspeed

Ask four pilots to describe airspeed and you will get four different answers, none of which is entirely correct. The difficulty with airspeed is that it must be measured, and, as we have seen, every measurement has some error. There are two types of error: bad assumptions (such as the flea-sized Sun) and blunders. Airspeed is subject to both. This ignores the question of units: knots or miles per hour? Knots are more natural in navigation.

True airspeed is the actual speed of an aircraft through the air. It is very difficult to measure, and really needs to be inferred. But it is definitely the number we want. Abbreviate it as *TAS*.

Airspeed is commonly measured with a device called a *pitot tube*. The pitot tube measures the difference between the static pressure (that is, the air pressure at the aircraft's position in space) and the ram pressure (that is, the pressure at the opening of the tube). Generally, the pitot tube is hooked up to some kind of spring or diaphragm arrangement that moves a pointer that indicates a number on a dial. The number so determined is the *indicated airspeed*. It is commonly abbreviated *IAS*.

A Rube Goldberg arrangement involving long tubes and springs is subject to some error. This is like a blunder. But there is also an assumption error: since the ram pressure is affected by the angle between the airflow and the pitot tube's inlet, it may be that the indication is grossly in error (indicating less airspeed as the angle becomes larger).[20]

To get a good reading, the instrument needs to be calibrated: the needle points to say, 100, when it should really point to 98. So, a cal-

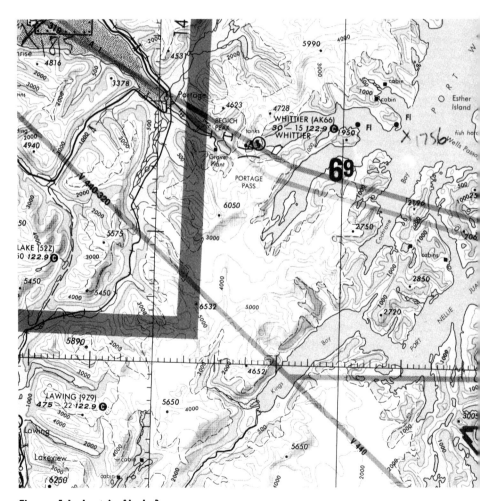

Figure 5-1. Lost in Alaska?

ibration table is prepared and included in the operating handbook. Now we have *calibrated airspeed*, which is abbreviated *CAS*.

Proper Use of Indicated and Calibrated Airspeed

Since IAS is directly related to the amount of air flowing over the lifting surfaces, it is very important to a pilot. For example, at a given weight and configuration (flaps and the like), an airplane will always take off at the same indicated airspeed.

The angle between the wing and the airflow is called the angle of attack. Most wings have a so-called critical angle of attack, above

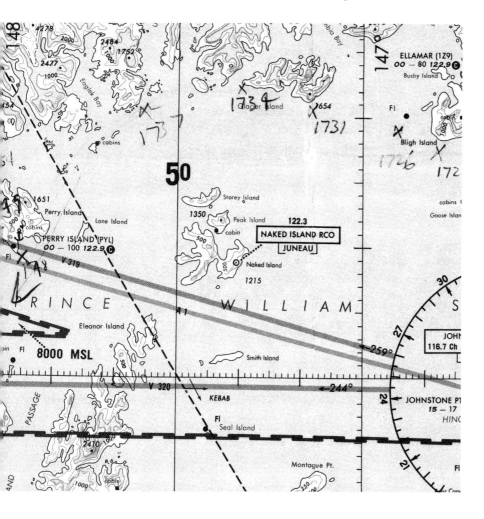

which the wing does not develop enough lift to support the airplane; this is generally at 18° or so for small airplanes. When the critical angle of attack is exceeded, the wing is *stalled*. The calibrated airspeed (in straight and level flight) at which the wing stalls is called the stall speed and is denoted, by Federal Law (FAR 1), v_s. The stall speed in the landing configuration is denoted v_{s_0}.

By long practice, the best calibrated airspeed for a landing approach is $1.3v_{s_0}$. In order to determine what the airspeed indicator should say, the pilot has to convert this back to IAS. For example, the wings-level stall speed of a Cessna 152 in the landing configuration with the most rearward permissible center of gravity is 41 KCAS (for knots of calibrated airspeed), but the IAS is 31 KIAS. (This much difference is not unusual.) An unwise pilot approaches at an indicated airspeed of 41 ≈ (1.3)(31), and is actually dangerously close to stalling (which leads, almost inevitably, to a significant loss of altitude). The wise pilot calculates the approach *calibrated* airspeed to be 53 ≈ (1.3)(41), which is an indicated airspeed of about 52. The manufacturer actually recommends a speed of 54 KIAS.

Groundspeed

There's one more speed to worry about: *groundspeed*, which is the speed that the airplane is traveling across the ground. An airplane with a true airspeed of 60 knots flying directly into a 60 knot wind has a groundspeed of zero; turn the airplane 180°, and the groundspeed is 120. We'll calculate the effect of wind in the next section, but for now note that the groundspeed is what the navigator really needs in order to know how long it will take to fly to the destination.[21] Boaters have a similar concept called the "speed over the bottom": currents mean that the speed of the boat through the water differs from the speed over the bottom.

Density Altitude and True Airspeed

Another problem with measuring TAS is that air is less dense at higher altitudes. This means that less pressure is exerted on the pitot tube, so IAS is typically a lot lower than TAS. Not only that, temperature has a measurable effect on air density: hot air is less dense than colder air, which is why hot air rises. So, on a hot day, the air is less dense and the indicated airspeed is even further reduced from the true air speed.

If we could determine the density of the air we are flying through, we could correctly determine TAS. But air density is hard to measure, especially while moving through it. Instead of measuring density, we use a *model* of the atmosphere called the *international standard atmosphere* or ISA.

The ISA is based on a standard pressure of 29.92″ of mercury and a standard temperature of 15° Celsius. In the lower part of the standard atmosphere, the temperature decreases by 2°C each thousand feet, and the pressure decreases by about 1″ with the same elevation change.

You can determine air density with a table: a given temperature and pressure imply a certain density. Using ISA, you can convert this density to an equivalent standard altitude; that is, you can say that the observed density is the same as at, say, 8500 feet at standard temperature. This is called *density altitude*. Density altitude determines true airspeed and, in general, airplane performance.

Estimating TAS from IAS

It seems that many pilots don't know their true airspeed. Ask them how fast their airplane is, and they'll say something like "On my last trip to Denver it was 200!" Most likely, this is a groundspeed derived from a LORAN, GPS, or map reading. Such a statement describes the wind on that day, not the airplane. The easiest way to estimate TAS may be to open the operating handbook and look it up (the operating handbook usually has corrections for temperature and pressure).

There is an empirical rule of thumb for low-altitude operations: TAS is about 2 percent more than CAS for each thousand feet above sea level. That is, if an airplane is at 8000′ above sea level, the TAS is about 16 percent more than the CAS.

This turns into a neat trick if the airspeed indicator has both knots and miles per hour on the same scale. At 7500 MSL, a common altitude for small airplanes, TAS is 15 percent more than IAS. But a nautical mile is 15 percent longer than a statute mile. So, at altitudes between about 5000 MSL and 10,000 MSL, the miles per hour scale shows approximate true airspeed.

Mach Number

Most jet pilots refer to speed as *Mach number*. This is the ratio of true airspeed to the speed of sound. The speed of sound is about 600 knots, but it varies with temperature, so Mach number alone is not good enough for navigation. In other words, on some days Mach 0.8 is 470 knots and on others Mach 0.8 is 490 knots.

Usually the approximation (M1.0 = 600 knots) is good enough because winds in the stratosphere (the layer of atmosphere above the troposphere, where we live) are generally light. Light means different things to different people, though: 30 knots is light for a jet, but a monster for my Taylorcraft. Most jets cruise in the stratosphere. Also, jets can carry enough fuel and there are enough alternates available, so the 10 knot difference is irrelevant. Jet pilots are more likely to talk about temperature than wind, since temperature often has more effect on performance. For precise navigation (such as when crossing an ocean with minimal fuel), the navigator needs to use the TAS.

One nice thing about Mach number and knots: since Mach 1 is about 10 miles per minute, Mach 0.5 is about 5 miles a minute, and Mach 0.8 is about 8 miles a minute.

The Wind Triangle

Determining groundspeed in flight is fairly easy. Measure the time it takes to fly between two fixes; call this T. On a chart, measure the distance between the fixes; call it D. Then the groundspeed is D/T. Easy, except for the need to convert units. If you go 9 miles in 4 minutes, your groundspeed (expressed in units per hour) is

$$\frac{9}{4} 60 = 135 \text{ knots}$$

since there are 60 minutes in each hour.

Predicting the Wind

One of the navigator's tasks is to determine, with some degree of reliability, how long a flight will take, and this means that the navigator must estimate the groundspeed *before* takeoff. The groundspeed depends on the *winds aloft* as well as on the TAS.

Winds aloft forecasts are available from various sources. In the United States, the National Weather Service launches weather balloons from selected locations twice a day and determines winds (and temperature and dewpoint) aloft by tracking them. This information is fed into computer models of the atmosphere. The computer models forecast winds aloft as well as surface phenomena. The accuracy of such forecasts is good in general terms but may be very poor in specifics. For example, in the Northern Hemisphere, air circulates in a clockwise manner around an area of high atmospheric pressure; so if the model forecasts a region of high atmospheric pressure in Missouri, you can predict northerly winds in southern Illinois. If the pressure gradient is high, then the windspeed will be high. But this is a far cry from saying that the wind at 9000′ will be from 332° at 36 knots.

Despite the vagaries, most preflight weather briefings in most countries include a forecast of the winds aloft at the flight's intended altitude, and the navigator has no choice but to base all of the planning on this forecast. A prudent navigator would look at possible alternatives in case the forecast is wrong.

What happens when the winds aloft forecast is wrong, but the navigator doesn't notice? See (Rickenbacker), who writes of a B-17 ditching in the Pacific during World War II. The first departure from Hawaii ended in a crash on the runway with no injuries, so the crew took a different airplane. The navigator's octant (an instrument like a sextant) had been damaged in the crash and gave inaccurate information, so he was unaware that the tailwind was much stronger than forecast. The airplane passed the destination, which was hidden by clouds.

The prudent navigator wants to know if the forecast winds are wrong even if fuel or course aren't issues, because of the implications about the general weather forecast. As a rule, in the continental United States, it is a very bad sign if the actual winds aloft are more southerly than the forecast winds aloft. Southerly winds bring moisture, which exacerbates all weather problems, so it is more likely that some kind of alternate plan will be necessary. The navigator who notes actual winds aloft and compares them with the forecast will notice potential weather difficulties sooner.

The FAA requires private pilot candidates to know how to get the winds aloft forecast and use it to make preflight estimates of groundspeed and wind correction angle. They must demonstrate this knowledge during the flight test. Some examiners make the candidate fly the predicted heading for the predicted time "under the hood," that is, on instruments. If the first checkpoint is reached, the candidate has demonstrated the ability to fly a heading by instruments, but the examiners tend to downgrade students who miss the checkpoint. They should really downgrade the forecaster who predicted the winds aloft. After all, they have seen the student do the computations and hold the heading; the only mistake would have been the forecaster's.

Wind Correction as a Vector Problem

What is the effect of the wind on the groundspeed? The navigator has four pieces of information that comprise one complete vector (the wind speed and direction) and pieces of two others. One is the aircraft's velocity vector through the air: the navigator knows the magnitude—the TAS—but not the direction. The other is the aircraft's velocity vector over the ground: in this case, the navigator knows the direction but not the magnitude, which is groundspeed.

The *wind triangle* relates these three vectors: wind velocity W, aircraft-through-the-air velocity A, and aircraft-over-the-ground velocity G. These are related: $A + W = G$. But because we don't know all of A or G, the solution requires more than simple vector subtraction.

To draw the triangle, proceed as follows. First, draw a line segment to represent the direction of G, that is, the desired course. The direction on the paper isn't all that important, since all of the other directions will be determined relative to the course line. Since the magnitude is unknown, you have to draw the line "long enough."

Pick a point on this line as the origin. Draw the wind line from the origin. This can be confusing, because wind direction is always the direction from which the air is moving. So draw this wind vector as if your pencil is being pushed over the page by the wind. You need to determine the difference between the wind direction and the course direction. This line, unlike the course line, can be drawn to scale since we know the forecast wind speed.

Finally, from the end of the wind vector, draw a segment whose length (to scale) is the TAS and that just touches the direction line. This is the complete triangle. It is illustrated in Figure 5-2.

The most common way to solve the problem is to draw the picture to scale and measure the results. For example, suppose you want to fly due north (360°) at 120 knots and the wind is forecast to be from 240° at 10 knots. The wind is a little bit behind you and from the left, and the difference between the wind direction and the course direction is 60°. Draw a line 10 units long from the origin to represent the wind speed. From the end of this, draw a segment 120 units long back to the course line, and measure. The *wind correction angle*—the angle between the vectors *A* and *G*—is about 5°, and the ground-speed is about 125 knots. A device called an E6-B computer can be used to solve this problem. Most student pilots own one.

Analytic Solution Using the Law of Sines

We can also solve this problem analytically, using the *law of sines*. A black box must use an analytic solution.

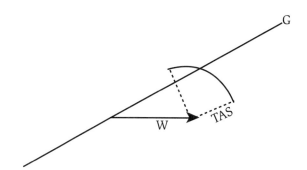

Figure 5-2. Course line, wind vector, and TAS.

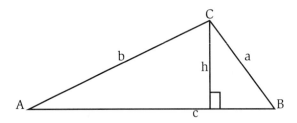

Figure 5-3. The law of sines.

The law of sines, like the law of cosines, relates sides and angles in all triangles. The general case is like this: Suppose that a triangle has sides of length a, b, and c, and that angle A is opposite the side of length a, that angle B is opposite the side of length b, and that angle C is opposite the side of length c, as in Figure 5-3. Then the law of sines says that

$$\frac{\sin A}{a} = \frac{\sin B}{b} = \frac{\sin C}{c}.$$

We can see that this is true by comparing any pair of sides, say the ones of length a and b. Drop a perpendicular from the angle C to the side of length c, as shown in Figure 5-3.

We can calculate the length of this perpendicular two ways: it is either $b \sin A$ or $a \sin B$. Setting these equal, we get the desired relation. This also works for a more obtuse triangle.

Now, go back to our wind triangle. Let WCA denote the wind correction angle, that is, the angle between the desired course and the actual direction you need to fly. It is the angle between vectors A and G; right now, it is an unknown. Let D be the angle between W and G; it is known. Then, using the law of sines, we have

$$\frac{\sin D}{|A|} = \frac{\sin WCA}{|W|}$$

so we conclude that

$$\sin WCA = \frac{|W| \sin D}{|A|}.$$

From the law of cosines, it follows that

$$|G|^2 = |A|^2 + |W|^2 - 2|A||W|\cos(180° - WCA - D).$$

An Example

For example, suppose again that the desired course is $360°$, that the TAS is 120 knots, and that the wind is from $240°$ at 10 knots. Then the angle D is $60°$, and $\sin 60° \approx .8660$. Thus, the sine of the wind correction angle is $10(.8660)/120 \approx 0.0722$, so the wind correction angle is about $4.1°$.

To determine the projected groundspeed, note that $180° - WCA - D \approx 115.9°$, whose cosine is about -0.4368, so

$$|G|^2 \approx 120^2 + 10^2 - (2)(10)(120)(20.4368)$$
$$\approx 15548$$

so

$$|G|^2 \approx 124.7.$$

Determining Winds Aloft

The reverse problem is of interest, too. If you know your ground-speed, track, true airspeed, and heading, you can determine the wind speed. Use the same symbols as before: A for air velocity, W for wind velocity, and G for velocity over the ground. It is always the case that $A + W = G$; in this case, since you know A and G, you can determine W by simple vector subtraction.

It is obvious that a headwind reduces groundspeed and thus increases fuel requirements, and it is obvious that a tailwind increases groundspeed and reduces fuel requirements. It's tempting to think that these even out over a round-trip, but that is incorrect. Here's a simple example. Imagine a round-trip of 200 NM in an airplane with a TAS of 100 knots. If there is no wind, the trip takes one hour each way, for a total time of 2 hours. What if there is a 20 knot headwind on the first leg, which becomes a 20 knot tailwind on the second? Now, the first leg is at a groundspeed of 80 knots, so it takes 1.25 hours, or 1 hour and 15 minutes. The second leg groundspeed is 120 knots, so this leg takes 50 minutes, for a total round-trip time of 2:05.

What about a 20 knot tailwind on the way out that decreases to a 15 knot headwind on the way back? Now, the first leg takes 50 minutes, while the second leg, at a groundspeed of 85 knots, takes 71 minutes, for a total time of 2:01.

It's also tempting to think that a direct crosswind has no effect. This is also incorrect, as an example will show. Suppose there is a 20 knot direct crosswind and that you are in a 100 knot airplane. Because of the crosswind, you will have a wind correction angle of about 12°. Your groundspeed will be a little less than 98 knots, not much of a loss, but not none.

Electronic Aids

There are many kinds of electronic aids available to pilots. We will discuss NDB (nondirectional beacon), VOR (VHF Omni Range), DME (distance measuring equipment), and various "random navigation" systems. (One early night navigation system was a series of coded light beacons defining routes. The state of Montana has kept a few of these working to help guide pilots through narrow mountain valleys, but most have disappeared.)

These systems were developed to help navigate in clouds or at night, and they are in many ways like navigating by pilotage and dead reckoning. Whether you determine your distance and bearing from a specific spot on the ground electronically or visually, you still know where you are.

But the electronic aids also offer something that makes them even more useful: a very accurate indication that you are on course. This is not something that was available for ships until recently through GPS. (Hutchins) wrote about watching a U.S. Navy crew navigate a large ship through a narrow channel. They had no on-course indication, and continually plotted their position using pilotage.

Even before GPS, aircraft navigators have long had a way to stay on course at night or in clouds by following electronic signals. If the needle on the instrument is in the proper position, the airplane is on course (in the old days, pilots had to interpret audio signals to stay "on the beam"). The most accurate ground signals come from the instrument landing system localizer, and since these are accurate to a few feet, they can be used to navigate safely very close to unseen obstacles.

Electronic Aids: Nondirectional Beacons

The simplest electronic aid to navigation is the *NDB* or *nondirectional beacon*. This is a simple radio station broadcasting a simple program: the Morse code identifier of the station.

To use an NDB, you need a receiver that can be tuned to the proper frequency and a rotating loop antenna (nowadays, the antenna neither rotates nor looks like a loop; it is manipulated electronically). The radio waves induce current in the loop, and this varies depending on the angle between the loop and the direction to the station. When the loop is aligned with the waves, the signal is strongest. When the loop is perpendicular to the waves, there is no signal.

A receiver that measures the signal strength while varying the direction of the (real or virtual) loop is called an *ADF*, or *automatic direction finder*. The typical cockpit display is a dial with a needle. The dial is supposed to show the airplane from above, and the needle points in the direction of the station. This is kind of an electronic Sumner line.

This information is useful. To find the Teterboro, New Jersey, airport, tune your ADF to WABC radio (770 kiloHertz). The WABC broadcast tower is less than 3 miles from the airport. Keeping the ADF needle pointing straight ahead will not lead you straight to the tower because varying wind will blow you off the direct course (this is called *homing*), but the needle will eventually lead you to the station and, hence, to the airport. Or you can apply wind correction and follow a straighter path to the tower and the airport.

ADF provides a simple means of finding a *LOP*. The key is that you have to know the airplane's heading, because the needle gives you only a *relative* bearing, that is, relative to the nose of the airplane.

Keep in mind that there are two common ways of measuring bearings: true bearings and magnetic bearings. We discussed these in the earlier section on the compass. Simple subtraction is all you need to convert between the two. Since aircraft use magnetic compasses, let's limit this discussion to magnetic bearings.

Suppose an aircraft is headed due north and the ADF indicates a bearing of 080° to the station. That is, if the aircraft were turned 080° to the right, the station would be on the nose. The relative bearing is 080°; the magnetic bearing is also 080°.

Now suppose that the aircraft heading is 045° and the relative bearing is still 080°. This still means that the aircraft must be turned 80° to the right to put the station on the nose, so now the magnetic bearing is 045° + 080°, or 125°.

If the relative bearing is, say, 310°, then the aircraft must be turned 310° to the right to put the station on the nose; it might be easier to turn 50° to the left, though. In any event, you add the relative bearing to the magnetic heading to get the magnetic bearing. Now, if the heading is 170° and the relative bearing is 310°, this gives you 170 + 310 = 480, and there is no such compass direction. But if you turn through 480°, you have the same effect as if you turn through 120° (because 480 − 360 = 120); so the magnetic bearing is really 120°. This is illustrated in Figure 5-4.

ADF and Avoiding Mental Arithmetic

The main objection pilots have to ADF is the amount of mental arithmetic involved. There really is very little need to do much, however. Since most heading indicators have marks at the 45° points, all you need to do is note where the ADF needle is relative to the nearest 45° point and mentally superimpose the needle on the heading indicator at the same distance from its corresponding 45° point. For example, suppose that the relative bearing is 120°; the nearest 45° point is 15° away (clockwise) at 135°; in other words, the needle is 15° counterclockwise from the 135° mark. Now, move your eye to the heading indicator, and read the magnetic bearing 15° counterclockwise from the mark on its case.

A nifty gadget, the *RMI*, or *radio magnetic indicator*, puts a rotating heading indicator (driven by a gyroscope) under the ADF needle. It's kind of like having the instrument superimposed already. You simply read the bearing under the head of the arrow.

In practice, NDB navigation is done over a very short range, and dead reckoning is all you need. At 120 knots (TAS) it takes 2.5 min-

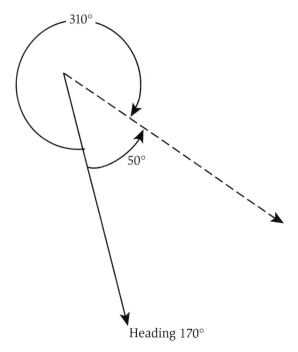

310°

50°

Heading 170°

Figure 5-4. Relative and magnetic bearings.

utes to go 5 miles, and a 20 knot crosswind component would push an aircraft only 0.8 miles off course in that time.

Advantages and Disadvantages of NDB Navigation

A more serious objection to ADF is that the signal isn't that accurate. The needle is subject to electronic interference and can swing maddeningly. You have to average out the swings to get an idea of the relative bearing.

The unsteadiness of the signal has led to several accidents with unusual chains of causality. A pilot was performing an instrument approach into LaGrande, Oregon, using an ADF. The airplane hit the top of a tree. Nobody was injured. During the accident investigation (see (NTSB)), the pilot said that the identifier was weak, but he was so accustomed to seeing the needle wander that he descended. The signal wasn't weak: the station was off the air for maintenance, and the wandering needle must have been much less steady than is usually the case.

So why use NDBs? They are inexpensive to build and operate. They require less land than a VOR. In fact, you can mount one on a truck. A retired airline captain told me of his company's first passenger-carrying flight into a certain Third World country. The aircraft was navigating using an NDB mounted on a truck. As the airplane approached, the truck driver realized that the truck needed gas and drove off to the fuel dump with the beacon on. Of course, the airplane went to the fuel dump, too. Since the crew was made up of prudent navigators, they did not descend, and nobody was hurt.

Problems with LOPs from Distant Beacons

There is a subtle problem with plotting NDB bearings. In Chapters 3 and 4, we calculated bearing to a destination and azimuth to a celestial body's *GP*, but it is a little more difficult to do this with an NDB bearing. For all of these, we know that the line of sight is a great circle, and we also know that the heading along a great-circle route is continuously changing. But for the first two cases, we plotted only a small portion of the great-circle route.

In plotting a long-range NDB bearing, there is a great temptation to plot the rhumb line instead, which is significantly different from the great circle. So, to plot an NDB bearing, you need to plot a kind of "line of constant bearing" from the NDB. To do this, plot the indicated bearing at each intermediate meridian. This shows you a curve made up of places on the Earth where the bearing observation is possible. This is not a straight line, even on a Lambert chart.

On a Mercator chart you will be able to plot only a rhumb line. A table of corrections is available for correcting the bearings you plot for more accuracy. See (Lyon).

Electronic Aids: VOR and DME

The most widely used electronic aid (even including GPS) is *VOR*, which stands for *VHF Omni Range*. Many pilots don't know where the name comes from. The "VHF" part is easy: the system depends on broadcasts in the VHF (very high frequency) part of the electromagnetic spectrum. The "Omni Range" part of the name differentiates it from an obsolete system called a radio range, which allowed pilots to navigate along only four predetermined courses. The VOR system allows the pilot to navigate over *any* course; hence "omni," a Latin prefix meaning "all."

The *V* in VOR is very important: radio signals in the VHF band can't turn corners, so you can receive a VOR station only when there is a clear line of sight between your antenna and the station. Two things can interfere: one is a mountain or a building between the receiver and the transmitter, and the other is the Earth itself.

NDB stations can be difficult to see from the air, but most VORs are circular white buildings with a large antenna sticking up from the center. There are other designs, but the round white buildings are the most common. They can be seen from several miles away and are useful in pilotage.

Many VOR stations also house something called *DME*, or distance measuring equipment. DME and VOR together enable a properly equipped aircraft to know both the bearing and the distance from the station; this is the same basic calculation we did with celestial observations. The DME usually works over a paired ultra high frequency (UHF) frequency and is subject to the same line-of-sight limitations as the VOR. Australia has its own VHF domestic DME system. DME is discussed in more detail below.

VOR Range

How far away is the horizon? Suppose you are at an altitude h (in thousands of feet). This is a simple right triangle whose vertices are you, the center of the Earth, and the point on the horizon. It's a right triangle because the tangent plane to a sphere is perpendicular to the radius. See Figure 5-5.

The hypotenuse has approximate length $3438 + h/6$; dividing the altitude by 6 converts the units to nautical miles (roughly). The line from the center of the Earth to the horizon has length 3438 nautical miles. Let D be the distance to the horizon. Then,

$$3438^2 + D^2 = (3438 + h/6)^2$$

so

$$D^2 = 1146h + h^2/36.$$

The last term is negligible, so

$$D \approx 34\sqrt{h} \ .$$

For example, when 10,000′ above the ground, $h = 10$, so the distance to the horizon is about 108 nautical miles.

Even at higher altitudes the range is limited to about 200 miles. See the *Airman's Information Manual* (FAA).

Using VOR

A VOR station (there are about 1000 in the United States alone) broadcasts two signals. One is nondirectional, and the other is a directional signal that works its way around the compass at a fixed rate. These are timed so that the nondirectional signal is broadcast as the directional one passes through magnetic north.[22]

The receiver in the aircraft measures the time difference between the reception of these two signals. This is proportional to the magnetic bearing from the station to the aircraft.

The cockpit instruments for VOR (Figure 5-6) are confusing for a lot of pilots. There are two ways to use them, corresponding to the

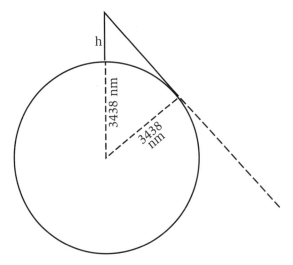

Figure 5-5. Distance to the horizon.

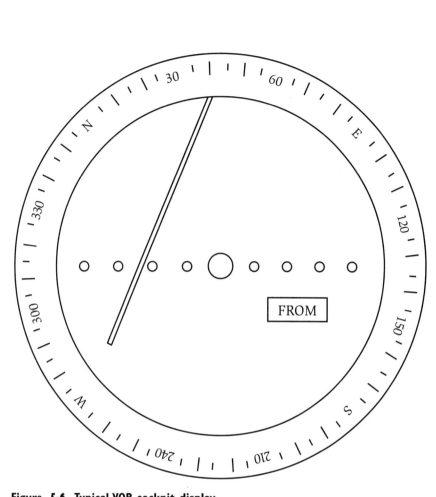

Figure 5-6. Typical VOR cockpit display.

two things a navigator does: one is to determine the aircraft's present position, and the other is to determine how to get to the destination. In either case, the VOR display consists of a needle called the course deviation indicator, or *CDI*; a *to/from indicator*; and a knob called an omni bearing selector or *OBS*.

The OBS is used to select a reference bearing. Suppose it is set to 100°. If the aircraft is generally east of the station, the to/from flag will say "from," while it will say "to" if you are generally west of the station.

The CDI will be to the left or right depending on whether the selected bearing would be to your left or right if your heading were

100. Your actual heading has no effect on the instrument, however. These are illustrated in Figure 5-7.

To determine how to get to the VOR station, you rotate the OBS until the CDI is centered with a "to" indication. This tells you your bearing to the station. If there were no wind, then that heading would take you to the station, but usually there is some wind. You could solve a wind triangle to determine the heading that will keep the CDI centered, but most people use a method called *bracketing* to determine the proper heading.

First, start by flying a heading that agrees with the OBS. If the wind blows you to the right of course, the needle will slip off to the left. Turn the aircraft 20° or so to the left until the needle centers again.

Now that you know that the wind is from the left, try a wind correction angle of, say 10° to the left. If this is too much, you will soon drift to the left of course, and the needle will move to the right. If it is too little, you will drift to the right of course, and the needle will move to the left. If it is just right, the needle will stay centered.

Suppose that the needle goes to the right. Then you know that 10° was too big a wind correction, so get back on course and then try 5°. If this is too big or small, you will begin to deviate again, but will know by the way you drift whether to try a smaller or larger angle. The next try would be between 5° and 10° if 5° weren't enough, but between 0° and 5° if 5° were too much.

It's natural that the CDI needle will move when the aircraft is close to the VOR because the radials are so close together; this is a lot like the way meridians converge at the poles. The prudent navigator

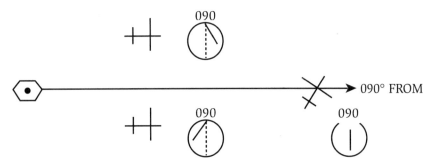

Figure 5-7. Course deviation indicator readings.

knows that the station is nearby and doesn't go into a lot of wild course corrections.

The CDI isn't an easy concept to grasp, but once understood it really is a nice way to navigate. It tells you whether your selected course is to the left or to the right, so you have an idea which way and how much to correct. When modern electronics such as LORAN and GPS first became available to aerial navigators, almost all of the manufacturers equipped their units with CDIs. These are a little different from the CDIs used by VOR receivers, since they show the distance from the course rather than the angle from the course (which is all that VOR can compute). While flying closer to a VOR station with a constant deflection of the CDI, you are getting closer to the course. With a distance-based CDI, you are staying the same distance from the course.

Many pilots fly long trips under VFR from VOR to VOR to VOR. In the United States, both the low- and high-altitude federal airway systems are based on VOR navigation.[23]

Problems with VOR

VOR is reliable, but there can be difficulties. First, in some installations, the course lines are subject to *scalloping*; they are not straight. An unaware navigator will spend a lot of effort chasing small deviations. If you have some idea of the aircraft's actual track, you will recognize that scalloping is occurring and keep flying the heading that has worked so far. If the needle deflection is persistent, however, a correction is necessary.

Near the station, the radials are very close together, and in most instances there will be a needle deflection to the right or left just before station passage. It is tempting to turn to try to keep the needle centered, but the navigator should be aware that the station is near and persist on the same heading that worked farther out.

Sometimes reflections from the propeller can cause the OBS needle to wander. A small change of RPM setting will prevent this.

It's important to check the identifier, too. Many stations use the same frequency. In this case, the stations should be far apart, but I was once in a King Air at 31,000′ when the two VOR receivers were tuned to the same frequency but were receiving different stations. (This was probably due to antenna placement.)

Line of Position from VOR

To determine where you are, turn the OBS until the CDI is in the center with a "from" indication. This then tells you what the direction is from the station to you. You can plot this directly on the chart to obtain an *LOP*. Two such LOPs, called a *cross bearing*, determine your location. See Figure 5-8.

The Local Nature of VOR

Since VOR uses the VHF portion of the spectrum, the signal is receivable only when there is an unobstructed line of sight between the antenna and the station. Generally, at medium altitudes, a VOR signal can be received for about 40 nautical miles; reception distance is lower at low altitude. The signal cannot go through mountains, and it can be bent by large objects such as buildings.

For the navigator, this short range has good points and bad points. It is not possible to tune Los Angeles VOR from upstate New York and follow the signal to Los Angeles; that's the bad news. The good news is that over short ranges it is possible to use the tangent plane approximation to the Earth's surface with sufficient accuracy. Remember that in this situation, each minute of latitude is one nau-

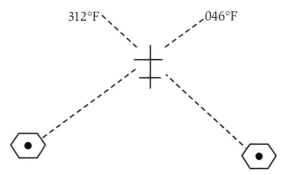

Figure 5-8. VOR lines of position.

tical mile long, and the length of a minute of longitude is approximately the cosine of the latitude.

DME

DME stands for *distance measuring equipment*. DME tells you your distance from the station and the to-from component of your groundspeed, and it is much more accurate than VOR at longer distances.

DME works in the UHF or ultra high frequency band; the same line-of-sight limitation on VOR reception also applies to DME. Each ground station is assigned a frequency. The airborne unit sends a discrete signal on this frequency; the ground station plays it back; and the airborne receiver determines distance from the station by the time it takes for the signal to come back (adjusted for processing time, which is a design parameter: 50 μsec). The signal comes back at the speed of light (about 186,000 statue miles per second or 582,000,000 knots) so the receiver only needs to do a simple distance-rate-time problem to determine the distance (after reducing the time by 50 μsec). The distance made good between two fixes provides a groundspeed estimate, but this measures only movement toward or away from the ground station. If you fly a perfect circle around a DME ground station, your distance won't change and your groundspeed will read zero.

Usually, but not always, DME stations are at VOR stations, and some are installed at runway ends to provide distance-to-go information for instrument approaches (this is often called "localizer DME," after the name of the azimuth transmission from the instrument landing system). A few are located at NDBs.[24] DME is also compatible with TACAN, a military navigational system, and many TACAN stations are located at VOR stations. In this case, it is called a VORTAC.

Accidents have been caused by using the wrong DME. Several years ago, a King Air crashed short of the runway in Salt Lake City because the pilot tuned the DME to Salt Lake VOR (which is not on the airport) rather than the localizer. The aircraft was 3 miles farther from the airport than the pilot thought.

The accuracy of DME is 0.5 miles or 3 percent, whichever is greater. How does this compare with VOR? When 60 miles from the station, the maximum DME error is 1.8 miles. This corresponds to the angle whose tangent is 1.8/60, that is, about 1.8°. But a VOR receiver may be as much as 6° in error and still be legal for navigation.

Slant Range

There is a problem with DME: the distance measured is a distance in space, not the distance along the surface. When the aircraft is far from the station, the difference is negligible, but it is significant if the aircraft's ground position is near the station. If you are 6000′ directly above the DME station, your ground distance is zero but your receiver says 1.0 NM. This is called *slant range error*.

People who design instrument approach procedures take this error into account and label the chart with the value your DME should display, not the ground distance. This error is particularly dangerous if the aircraft has to fly directly over the VOR/DME station, because you must account for the error twice.

Another problem with passing right over the station is that the distance first counts down to a minimum (not zero) and then starts to count up again. Often a descent starts at a DME fix; for example, the descent might start at DME 3.0. A navigator not keeping track of position might start the descent at the wrong 3.0 DME fix and hit something.

Electronic Aids: FMS

The *FMS*, or flight management system, was one of the first applications of computing to aircraft navigation. Before FMS, all navigation was done by hand. That is, a navigator (or pilot acting as a navigator) had to plot compass bearings, take celestial fixes, measure drift, estimate groundspeed, and tune radios by hand, on paper.

Flight management systems use sensors to estimate position. FMSs have been based on inertial navigation, GPS, LORAN, Omega, and VOR/DME. In some sense, how the FMS determines your position makes no difference in how you navigate with it. In all cases, you tell the unit where you are starting and what route you want to fly. In flight, it determines or estimates your position and determines your distance and bearing to the next waypoint, using the same equations we used to calculate altitude and azimuth in Chapter 3. This section is about how the FMS estimates the current position; the next is about how to use the FMS.

One nice thing about the FMS: most have a *moving map* display. Over the years, more and more pilots rely on the moving map rather than using paper charts. The moving map is wonderful when everything is working: it usually shows nearby navaids and airports, and some even include shorelines and other ground features.[25]

I was once a guest in the cockpit jumpseat of a Boeing 757 going from Denver to Boston, sitting behind and above the captain, when the moving map wasn't good enough. We were over western Illinois talking happily about the things pilots talk about when over western Illinois. A severe weather system was moving into Indiana, and soon there were too many airplanes trying to occupy too little space in Indianapolis Center's airspace. The Indianapolis controllers, in a justifiable panic, closed their airspace to incoming traffic. (This applied to IFR traffic only.) Suddenly our conversation was interrupted by a radio call telling us to hold at the LOAMY intersection. We were going 9 miles a minute, and LOAMY was 12 miles ahead, so the captain and first officer had less than 80 seconds to configure the jet for the hold. There are two aspects to this: slowing down, and programming the FMS to do the hold.[26] Eighty seconds isn't enough time for this, and hands were flying everywhere around the cockpit. Not mine, of course: I was a guest; I was not qualified to fly a Boeing 757; and I couldn't reach anything anyway. Mine were folded in my lap, where a guest's hands belong.

But the FMS didn't show LOAMY or the holding pattern there. I was no longer a guest: the captain grabbed a chart and handed it to me over his shoulder, saying "Find that intersection!" I reached forward to grab it and, by luck, my thumb was right on the intersection. I yelled "Got it!" right away. (Oddly, the controller's holding instructions disagreed with the chart, and there were a few radio calls to get this right.)

Inertial Navigation

INS, or inertial navigation system, measures accelerations, which enables it to determine your velocity *in space.* This in turn allows an estimate of current position *in space.* It's like doing dead reckoning 60 times a minute. But there's a catch: this scheme estimates the aircraft's position in space, so an an Earth model (usually WGS-84) is used to estimate your latitude and longitude.

One big problem with inertial systems is that the Earth is spinning. Suppose you turn one on and sit perfectly still on the ground. An hour later, you are at the same point on the Earth's surface but you have moved in space. In other words, INS determines your position with respect to the celestial sphere, not the Earth. The design must account for this and subtract the Earth's known (or estimated) movement.

Another problem is that there is always a little error in the estimate of the acceleration, which leads to a little error in the velocity estimate, which leads to a little error in the position estimate. By using carefully calibrated accelerometers and lots of precision in the calculations, these errors can be reduced, but not eliminated. This accumulated error is called *drift*; to be usable, the drift is supposed to be less than 0.8 NM/hr. This is excellent for a short flight, but I have been a passenger on routes like London to Los Angeles and New York to Tokyo. These flights last 12 hours or more, so there is the possibility that the inertial position is nearly 10 miles in error.

In practice, the navigator can reset the drift to zero by telling the INS that it is at a certain point. Typically, an aircraft starting across an ocean crosses the coast at a VOR. As the VOR is crossed, the navigator hits the "hold" button, which displays the system's estimate of the current latitude and longitude. The navigator compares this with the VOR's known latitude and longitude and corrects the INS position. Modern systems use GPS to update the INS position; this way, if one fails, the other takes over.

Another source of error is called Schuler drift. This is a periodic error with a period of about 84 minutes (the period of a pendulum whose length is the diameter of the Earth). This error is small but noticeable, and makes it appear falsely that the aircraft is off course. Navigators using INS tend to attribute small errors to the Schuler effect and ignore them. If this is the true cause, the error will vanish; if the error persists, a course change is called for.

Inertial navigation replaced *Doppler navigation*, which used radar to estimate groundspeed and drift. Smooth bodies of water don't reflect radar as well as land or choppy water (I like to make radar pictures of large lakes enroute), so Doppler doesn't work well over smooth oceans. The Pacific Ocean is named for its smoothness (the name's root is the same as for "pacifist"), so Doppler wasn't always reliable there. Doppler replaced celestial, at least partially.

When INS became operational, airline captains reported that they could now actually see aircraft reporting the same position, and that aircraft flying the opposite direction on the same route now passed directly overhead (see *Flying*, December 1987). This led to a silly problem: some autopilots are programmed to disconnect themselves if the aircraft descends to within 2000′ of the ground unless commanded otherwise. Now there were autopilot disconnects at high altitude in mid-ocean. Why? The aircraft had passed directly over another aircraft, and the radar altimeter had interpreted the radar return from the aircraft below as the ground.

Some FMS designs allow automatic updating: periodically the INS position is updated based on fixes from ground-based navaids. This can lead to some navigator discomfort while crossing an ocean. While over the ocean, the system receives no VOR or DME signals, so the INS position drifts. But, as the aircraft gets closer to shore it starts to receive signals, and using these, it computes a position (see below for the method) and updates the INS position. If the navigator happens to be watching the moving map the first time the position is updated, he or she will see it shift to reflect the updated position. This is called *map shift*. Experienced crews anticipate the minor course change as the airplane gets close to shore.[27]

Estimating DME-DME Positions

Many FMS systems use VOR and DME to determine position. This is done with an important mathematical technique called *iteration*. An iterative technique starts with a guess, compares what the guess implies with reality, and improves the guess. The Saint-Hillaire method is iterative. In each case, the analytical solution is almost impossible to derive, and the extra accuracy isn't needed.

For VOR/DME updating, the FMS tunes VOR and DME stations automatically and uses the information it receives to update the position estimate. Since DME is more accurate than VOR, the best estimates come from two DME stations. Here's how it goes: the FMS knows the positions of DME A and DME B, and estimates the current aircraft position to be p_0. It then calculates the distance from p_0 to each station. If the calculation agrees with the observed DME measurements, the guess was correct.[28] The difference between the actual distance and the distance calculated from the estimated position is called the *residual*.

If the calculated and observed measurements differ, the estimated position is moved toward or away from each station to get a new estimate p_1. The distances from p_1 are computed and compared, and the iterative process continues. See Figure 5-9.

When flying with an FMS with moving map doing DME-DME updating, the screen sometimes appears to jiggle, and sometimes you see the symbol for a navaid skitter across the screen before stopping at its approximate relative position. What you're seeing is the update process in action: the moving map shows each estimate in turn.

One method the system can use to obtain the initial guess is to calculate a dead-reckoning position based on the last estimate. This works best when the aircraft is flying a constant heading; the FMS takes the estimate position, estimated groundspeed, and estimated track (all known from previous calculations) and calculates the DR. If you don't turn, this will be correct, and there is no need to refine the estimate. If you do turn, the error in position after one second (a typical update time) is minimal, so the estimate may need only one or two revisions. Or once the FMS realizes that a turn has started, it might assume that the turn is continuing and use this as the basis of the DR.

Another Iterative Method: Square Roots

A very striking and sometimes useful illustration of iteration is the following method for estimating square roots; it is a special case of

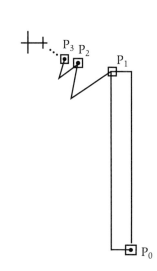

Figure 5-9. DME-DME updating.

Newton's method. The process is called "guess-divide-average" and is well illustrated by estimating the square root of 4 (the demonstration is more convincing when you know how the calculation should come out).

First, make an initial guess that $1 = \sqrt{4}$. Of course this is wrong, so we divide 4 by 1 (getting 4) and average this with our original guess to get $(1 + 4)/2 = 2.5$. Our next guess is therefore 2.5. Divide 4 by 2.5; you get 1.6. The average of 2.5 and 1.6 is 2.05. Divide 4 by 2.05, getting about 1.9512; the average of 2.05 and 1.9512 is 2.0006. As you can see, we are very close to the exact result, namely 2.

This algorithm still works if you begin with a different guess. Suppose you guess that $\sqrt{4} = 10$. Then $4/10 = 0.4$, and the average of 10 and 0.4 is 5.2. Now divide: $4/5.2 = .76923$. Average this with 5.2 to get 2.9846. Divide: $4/2.9846 = 1.3402$. Average this with 2.9846 to get

2.1624. And so on. If you make a bad guess it adds a round or two to the calculation, but the result still approaches 2 very rapidly.

Other Issues with Iterative Methods

Guess-divide-average has an advantage over DME-DME updating: the size of the next step is determined for you. For DME-DME updating, you need to decide how far to move the estimate each time. If your computed distance is too large, you need to move toward the station, but how far? One mile? A proportion of the difference? If you move the estimate too far you will pass the true position, and you may find yourself chasing your tail. If you move too little, it will take a lot of moves to get a final estimate. The answers to these questions are hard to pin down because they vary from problem to problem; this is one of the design tasks avionics engineers face.

One possible algorithm starts as follows: let the DME locations be latitude L_i and longitude λ_i, where i can be 1 or 2. Locate p_0, the first guess, at the smaller longitude and smaller latitude. For example, if $L_1 = 30$ and $L_2 = 31$, the latitude of the assumed position would be 30; if $\lambda_1 = 80$ and $\lambda_2 = 79$, the longitude of the assumed position would be 79.

During each round, compare the calculated distances with the observed distances to the stations. If the calculated distance is too large, move the assumed position toward the station by the amount of the residual; if the calculated distance is too small, move the assumed position away from the station by the amount of the residual. You now have a new estimated position.

The residuals should get smaller and smaller; when they are small enough, you are done. In some cases, you might find that the residuals do not become smaller. This is a sign that something is wrong, and the computer should issue some kind of warning.

Electronic Aids: Using GPS, LORAN, and Other RNAV Systems

The theory of GPS was covered in Chapter 4, so here we will concentrate on how it is used. The same methods apply to all of the "automagic" black boxes in use, including LORAN and FMS systems.[29] There will be no discussion of the regulatory aspects of using these boxes; the regulations are changing very quickly, and anything on paper is bound to be obsolete before you read it. In this section, *GPS* means any of the electronic long-range navigation systems.

Sometimes You Can't Go Direct

One common usage of GPS, especially under VFR, is to fly direct to the destination. The great-circle distance from Pocatello to Seattle is 492 nautical miles. The airway distance is 513 nautical miles. A pilot limited to short-range methods would have to fly the longer airway route, but a one with GPS can consider the direct route. Unfortunately, the direct route passes very close to Mount Rainier, so many airplanes have to use a different, longer, route.

A similar problem occurs on the direct route between Salt Lake City and Los Angeles. It's shorter than the airways route, but the direct route passes through a lot of "special use" (that is, military) airspace, and most of the time the airways route must be flown.

A GPS may or may not have all of this information in its database, and the database may or may not be up-to-date. So you still need the paper charts.

The prudent navigator needs to keep in mind that GPS can fail due to terrain blocking, jamming, electrical failures, interference from onboard radios, change of political regime, or (my favorite) pilot or navigator error.

Interpreting Track

One of the most useful things that a GPS can give a pilot is the airplane's *track*, that is, the actual direction the airplane is traveling over the ground. This is the same direction that a wind triangle would determine if the actual winds aloft were known. If the track to your destination is the same as the bearing to your destination, you are headed straight there. Or you can fly parallel to a given course line at a constant distance by having your track match the charted bearing.

GPS also gives you an accurate groundspeed. Combining this with the track gives you complete knowledge of the vector *G* used in the wind triangle. Since you also know your heading and true airspeed, you have complete knowledge of the vector A. Since $A + W = G$, you can solve for W, the wind vector, as $G - A$.

The more capable FMS systems have inputs from the compass and the airspeed indicator,[30] so they have constant knowledge of both A and G and are thus able to display the actual wind vector continually.

GPS and IFR Navigation

When GPS and other black boxes first became reliable enough to use for IFR flight, there was some question about how to use them. The most common method to date has been as an *overlay*. For example, many airports have instrument letdown procedures based on a certain VOR or NDB. See Figure 5-10.

A pilot with a properly equipped and installed GPS can fly this procedure without tuning a VOR at all. The black box knows where the VOR is, after all, and can determine your bearing from the station's location. The advantage is that the VOR-based procedure has been designed so that the airplane is always at a safe altitude (remember, these procedures are used in situations of low ceilings or low visibility). Knowing that, say, 4100′ MSL is a safe altitude at a certain point is independent of how one determines that one is at that point.

In the United States, the FAA is also designing completely new procedures based on satellite or FMS navigation alone; these are called RNAV procedures. Some of the procedures based on VOR have altitude restrictions because of the problem of maintaining line of sight to the station, but satellite-based RNAV procedures will be immune from this problem and may allow aircraft to descend to much lower altitudes.[31]

GPS Waypoints

Most black boxes allow the user to define *waypoints* that define the route. A waypoint is just the assignment of a name to a certain latitude and longitude. If the waypoint is at the location of a VOR, for example, you could use the name of the VOR. The same applies to NDBs, airports, and charted intersections (that is, points where airways meet or change).

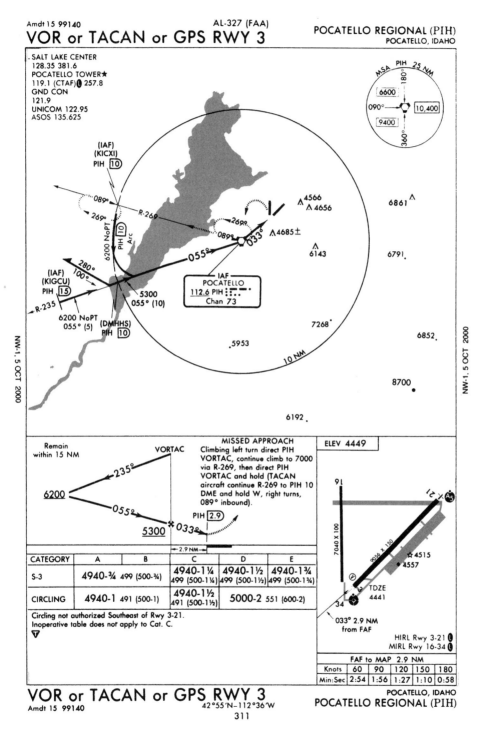

Figure 5-10. Typical VOR approach.

IFR procedures involve many waypoints that do not already have names but that can be determined from known waypoints. For example, consider the VOR approach chart in Figure 5-10. In following such a procedure, the aircraft follows certain courses at certain altitudes until reaching the "missed approach point" (MAP). If the airport is in sight, the pilot can go ahead and land; if not, the pilot needs to climb away from the ground and come up with another plan.

Most MAPs aren't named; the most typical situation is the one in Figure 5-10, where the MAP is 033° degrees from the Pocatello VOR at a distance of 2.9 nautical miles. So in order to use the MAP as a waypoint, we need to determine its latitude and longitude.[32]

We know the latitude and longitude of the Pocatello VOR: the latitude is 42° 52.2′ North, and the longitude is 112° 39.1′ West.

As a first approximation, we can work in the tangent plane to the Earth at the VOR. One minute of latitude is, of course, one nautical mile, while one minute of longitude is $\cos(42° 52.2′) \approx 0.7329$ nautical miles. The magnetic bearing of 033° corresponds to a true bearing of 048°, since the variation is 15° East.

Notice that the change in latitude is 2.9 sin(048°), that is, about 2.2 nautical miles. So the latitude of the MAP is about 42° 54.4′ North. Similarly, the MAP is 2.9 cos(048°) or approximately 1.9 miles east of the VOR. Since one minute of longitude is about 0.7329 miles, this represents about 2.6 minutes of longitude which must be subtracted from 112° 39.1′, leaving us with 112° 36.5′ as the longitude of the MAP. This kind of approximation is often good enough to use if the distance is short, say, less than 10 nautical miles.

Analytic Methods

More generally, though, we need to use the geometry of the sphere to determine the latitude and longitude of the MAP. The general problem is this: given a point, say A, at latitude L and longitude λ,[33] what is the latitude M and the longitude μ[34] of the point B at bearing θ and distance ρ?

In practice, this problem is solved iteratively, using a method like that for DME-DME fixes. You start with a guess for the MAP coordinates, and calculate the distance and bearing from the VOR to the guess. If the bearing is off, move your guess left or right. If the distance is off, move your guess toward or away from the station.

Analytically, the first method that you'd try to use in solving this is to use the formulas from Chapter 3: calculate the distance from A to B and set it equal to ρ, and calculate the bearing between them and set it equal to θ. Don't try it; it's a mess!

It's better to look at the geometry of the problem. First, all the points at distance ρ from A lie on a circle, and this circle lies in a plane. See Figure 5-11. Similarly, the points at bearing θ from A lie in a different plane, as shown in Figure 5-12.

These two planes intersect in a line, and one of the spots where that line intersects the sphere is the missed approach point. If we can determine its Cartesian coordinates, we can convert them back into the latitude and longitude of B.

We could determine the equations of these two planes, but the calculation is messy (again), so we will restrict ourselves to a special case and then determine how to proceed from the special case to the general case.

The first guess for a special case on the sphere is the North Pole, but bearings aren't defined there.[35] Instead, we will use the point on the equator with Cartesian coordinates $(R, 0, 0)$, where R is the radius of the planet. Of course this is the point at latitude 0° North and longitude 0° West.

The plane containing the circle of points at distance ρ has normal $(R, 0, 0)$, or better yet $(1, 0, 0)$. Let L be the distance from the origin to

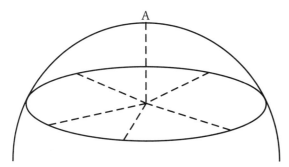

Figure 5-11. Points at a fixed distance from A.

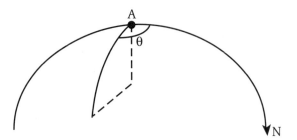

Figure 5-12. Points at a fixed bearing from A.

this plane. Then the plane (Figure 5-13) containing this circle has the equation

$$x = L.$$

To determine L, notice that the radian measure of the angle subtended by an arc of length ρ is ρ/R, and, from the Figure 5-13,

$$L = R\cos(\rho/R).$$

The second plane makes the dihedral angle θ with the plane $Y = 0$. The normal to this plane is therefore $(0, -\sin\theta, \cos\theta)$. Since it passes

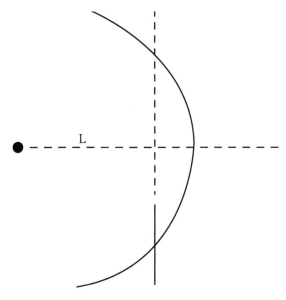

Figure 5-13. Plane at distance L.

through the origin the equation is

$$-\sin\theta Y + \cos\theta Z = 0.$$

These planes intersect along the line through $(L,0,0)$ parallel to $(0,1,\tan\theta)$. That is, for each point on the line there is a value of t such that the coordinates of the point are $(L,t,t\tan\theta)$. The point we want corresponds to the t-value where this vector meets the sphere, that is, where $L^2 + t^2 + t^2\tan^2\theta = 1$. Collecting all of the t-terms gives us

$$L^2 + t^2(1 + \tan^2\theta) = 1.$$

Remember that $\sin^2\theta + \cos^2\theta$ is always equal to 1. Away from places where $\cos\theta$ is zero, we can divide through by $\cos^2\theta$ and obtain the identity

$$1 + \tan^2\theta = \sec^2\theta$$

which holds for almost all θ. Using this, the equation for t becomes

$$t^2 = \frac{1 - L^2}{\sec^2\theta}.$$

Since $\sec\theta = 1/(\cos\theta)$, we get

$$t^2 = 1 - L^2\cos^2\theta.$$

You need to take a square root to determine t from the formula above. Positive real numbers have *two* square roots. The calculation has been set up so the *positive* square root is the one needed. Some common sense would also be helpful here.

Now, how do you translate this to the general situation? Remember, the calculation above finds a point a distance ρ and bearing θ from the point at latitude 0° North and longitude 0° West. In the case of any known point on the equator, it gives us the *difference* between the longitude of the known point and the longitude of the waypoint. Just rotate the solution to the proper longitude.

It's a little harder if the point is off the equator. The easiest thing to do is to get a solution at the proper longitude and then rotate the solution to the proper latitude. The axis of rotation goes through the center of the Earth 0 and is perpendicular to the line from 0 to the point in question. The Cartesian coordinates of a point on the equator with longitude λ are $(\cos\lambda, \sin\lambda, 0)$; the normal line is therefore $(-\sin\lambda, \cos\lambda, 0)$. Rotate about this axis until the latitude of the known point is reached.

Miscellaneous Topics

This section contains a few navigational tricks that may be of interest.

Turn the Airplane and Use the Compass

Sometimes it seems difficult to get an accurate azimuth line of position from a prominent object such as a mountain: how can you determine the compass bearing exactly? One seldom mentioned method is to actually turn the airplane so the object is directly ahead, and read the heading off the compass. Apply compass deviation and magnetic variation to get the reciprocal of the true bearing from the object. The few seconds off course are not enough to have any significant negative effect.

Danger Bearing

At night, on instruments, in low visibility, or in potentially confusing locations (such as, the Los Angeles basin), it is important to plan *in advance* how to stay away from areas of danger. Danger may come in the form of a mountain or in the more subtle form of some kind of restriction to access to the airspace.

You can do this by plotting a *danger bearing* from a nearby VOR or NDB. For example, I regularly use a danger bearing off the Ogden VOR in order to avoid accidentally entering the Class B (clearance required) airspace near Salt Lake City. If the airplane is west of the 360° radial, then I cannot possibly be in the Class B airspace. By tuning the VOR to Ogden and setting the OBS to 180 while traveling southbound, I make sure I am safe by keeping the CDI to the left. See Figure 5-14.

Danger bearings have long been used at sea. In fact, sometimes shore lights are built in certain locations to provide a visual reference.

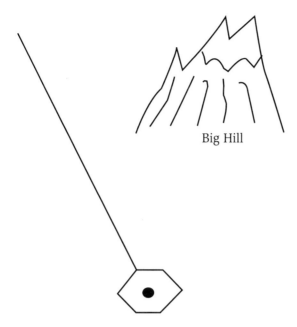

Big Hill

Figure 5-14. Danger bearing.

By staying lined up with two known or charted shore lights, a ship can avoid shoals or other dangers.

Point of No Return

For long flights over water or in other areas where there are not enough suitable airports, it can be important to calculate a *point of no return*. If something goes wrong before reaching this point, you return to the departure airport, but if something goes wrong later than this point, you continue to the destination. The key idea is to know how much time it would take to go to the nearest safe point, so the point of no return is calculated based on your estimated or, preferably, actual headwind or tailwind component. The outbound groundspeed will generally be different from the return groundspeed because a tailwind in one direction is a headwind when you turn around.

Suppose that the total distance is d, the true airspeed is T, and the tailwind component is W (which may be negative).[36] Let p be the distance to the point of no return; it is the unknown. See Figure 5-15.

$$\underline{\qquad\qquad\text{p}\qquad\qquad | \qquad\qquad \text{d}-\text{p}\qquad\qquad}$$

Figure 5-15. Geometry of the point of no return.

The time to travel from the point of no return back to the point of origin is

$$\frac{p}{T+W}$$

while the time to travel from the point of no return to the destination is

$$\frac{d-p}{T-W}.$$

Notice that the sign in front of W is different for the two directions, since an eastbound tailwind is a westbound headwind.

Setting these two times equal, we see that $p(T - W) = (d - p)(T + W)$. Bringing all of the p terms to the right gives us

$$p(T - W + T + W) = d(T + W)$$

so

$$p = d\frac{T+W}{2T}.$$

Double the Angle off the Bow

Lacking DME or some other reliable means of determining the distance to a destination can be uncomfortable. In this case, a very old technique called *doubling the angle off the bow* can be used if you are receiving a VOR or NDB. You alter course 5° to one side or the other, and wait for the bearing to change the same amount. See Figure 5-16.

When the bearing difference is the same as the course difference, you are halfway to the destination. This is because the triangle is an isosceles triangle.

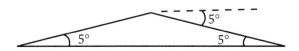

Figure 5-16. Doubling the angle off the bow.

This can be generalized using the law of sines. Suppose that you turn 5° left and wait until the bearing is only 1° to the right. See Figure 5-17.

The distance left to go L' is related to the distance traveled L by

$$\frac{\sin 1°}{L} = \frac{\sin 5°}{L}.$$

If α is a small angle then $\sin \alpha \approx \alpha/60$. Therefore L' is approximately $5L$.

Figure 5-17. Variation on doubling the angle.

NOTES

1. Many of these units, especially those meant for smaller cockpits where space is at a premium, use some combination of knobs and buttons rather than a keyboard.

2. In the United States, a pilot is allowed to use an out-of-date database if every fix is compared against a current paper source.

3. We ran into another hiker who was really lost and tried to convince him to follow us to the trail. For some reason he kept veering away from north heading, which should have been easy to follow by keeping the afternoon Sun on his left.

4. This is *not* legal in some countries, notably Canada.

5. A standard saying about IFR: one peek at the ground is worth a thousand cross-checks.

6. *Aviation Week and Space Technology*, September 22, 1997.

7. In case you want to be a Canadian commercial pilot, it is rumored that the answer to this question is B. If that's too hard to remember, you can use the methods in Chapter 3.

8. See (Wohl).

9. I have personally verified that these are reversed in the Southern Hemisphere.

10. More advanced navigational systems, including some GPS receivers, use a mathematical model of the Earth's magnetic field to convert true bearings to magnetic.

11. See (Sobel).

12. Recall the discussion of great-circle routes in Chapter 3. A detailed example is presented in Chapter 6.

13. The FAA likes U.S. pilots to know this; it appears on their knowledge tests.

14. The U.S. government agency responsible for producing civilian charts.

15. This is, of course, 7 miles a minute.

16. My sister is a historian and wrote to me from overseas saying that she needed a map with topographic information to help with her research. I sent her an ONC. These are easily available and inexpensive. We did not realize they were illegal in the country in question until she was visited by the secret police.

17. The Jeppesen Company publishes IFR charts for the whole world using a slightly different format than the local government publishers. The "Jeppesen versus government" debate is one of the most persistent discussion topics in aviation.

18. (Gann) wrote that pilots used this an inside joke when asked what town they were over, since they often didn't know the actual name.

19. This is one of the reasons the pilot and navigational functions are split up in aircraft with more than one crew member.

20. In some aircraft, flaps or landing gear can change the airflow sufficiently to change the airspeed indication.

21. There is yet another airspeed, called *equivalent airspeed*, which takes compressibility into consideration, but this applies only at speeds above 200 knots or so.

22. At least in principle. For practical reasons, there may be some difference between the VOR's north and magnetic north. This can be caused by changes in magnetic variation, for example. And at high latitudes, where the magnetic compass is unreliable, many VORs are oriented to true north.

23. A few exceptions are some low altitude airways determined by NDBs, especially in Alaska. These are called *low-frequency airways*. In some cases, a pilot has a choice between two airways that start and end at the same point, one defined by NDB and one defined by VOR. Sometimes the minimum altitude is much lower for the low-frequency airway, since NDBs don't suffer from the line-of-sight limitations that reduce VOR range. The airplane must be at a higher altitude to have adequate VOR reception, although a lower altitude would keep it from hitting any mountains. In some cases, this is advantageous: I once chose to fly a low-frequency airway in Alaska because to climb high enough to fly the parallel VOR airway would have put us into severe icing conditions.

24. For example, Hailey, Idaho, has an NDB/DME, which is used for flying instrument approaches to the airport in a narrow valley.

25. An extra-fancy version of the moving map includes a terrain warning system such as *enhanced ground proximity warning system*, or *EGPWS*. This uses a digital terrain database to show mountains and valleys ahead.

26. Readers who have had instrument training might recall how hard it is to program yourself to hold.

27. Another form of map shift, mentioned in Chapter 4, happens when the destination country uses a different datum than the aircraft. Fewer and fewer countries do this, but the practice continues.

28. In practice, one only demands that the observed and calculated distances be close. Since DME may be in error by 0.5 NM or 3 percent, you can't demand that the agreement be any closer.

29. Another system called Omega was decommissioned in September 1997.

30. More likely, from the *air data computer*, which determines the density of the ambient air and thus the true airspeed.

31. At this writing, the designers of RNAV approaches are actually requiring higher altitudes even though the obstacle clearance area is smaller.

32. As things stand now, this information must be in the database before a pilot can use it for an instrument letdown.
33. The Greek letter *lambda*.
34. The Greek letter *mu*.
35. A bear walks one mile due south, then one mile due west, then one mile due north, and ends up where it started. What color is the bear?
36. There are many considerations in deciding which way to go if a diversion is necessary. These are discussed in more detail in Chapter 6. Some of them need to use a different true airspeed.

Navigation Planning

Introduction

A prudent navigator's job begins long before the journey begins. One of the navigator's tasks is to choose a route and plan alternative courses of action. Planning can appear to be very easy, especially while on a familiar route in a familiar aircraft. To take my Taylorcraft the 44 miles from Pocatello to Idaho Falls, I can follow the interstate highway; it will take about 45 minutes; and I can carry more than 3 hours' fuel. If something goes wrong, I can land in one of hundreds of potato fields (and my insurance will pay for the ruined spuds). In this case the planning has been done, but it is remembered rather than done anew.

The simple example of a 44 NM journey deals with the main issues. The first purpose of flight planning is to not run out of fuel, and the second purpose is to not get lost. Navigation management systems also make planning seem simple: just look up the destination's identifier and enter it into the FMS. We have already discussed some of the problems with this approach.

In this chapter, we will do some detailed flight planning in a situation that requires more care than a simple trip to the next town. Our example aircraft is the Cessna 414A, a pressurized twin-engine airplane that I fly as an air ambulance pilot. Although performance is not one of this book's topics, we will discuss it in enough detail to ask the question: can we take this airplane from North America to Europe? We will consider the stock airplane, with none of the after-market enhancements such as better engines and vortex generators that improve performance in demonstrable ways.

I have never crossed an ocean in a Cessna 414, although I do try to plan my longer flights with this level of care, especially since my home airport is in the Rockies and alternatives are few. The same principles apply to oceanic, remote, or unpopulated areas. The ideas in this chapter are based on those in (Lyon). Current charts and certifications are needed for an actual flight.

There isn't much geometry in flight planning: we need to consider the effect of wind and terrain, and we need to know how to calculate bearing and distance between waypoints. The main tool is "distance equals rate times time" and persistence in applying this rule systematically.

Performance Basics

The purpose of flight planning is to determine whether you have enough fuel to complete the flight. Fuel consumption is an important performance consideration in all aircraft.

Depending on the year of manufacture and the rules under which it is built, an airplane has either a pilot's operating handbook (POH) or an aircraft flight manual (AFM). This book includes performance data necessary for flight planning. These data are derived during factory flight test, and a prudent navigator in a working airplane tends to expect a little bit less. As a minimum, the following data are included:

- Takeoff distance

- Rate of climb

- Time, fuel, and distance to climb

- Cruise performance

- Time, fuel, and distance to descend

- Landing distance.

The climb, cruise, and descent tables include speeds (assuming no wind) and fuel consumption. Since airplane performance is determined by air density, most of the tables give some way to convert pressure altitude and temperature to density altitude. In this section, we are assuming a so-called standard day, but in actual planning you would use the appropriate density altitude for that day's weather conditions. Pilots take this seriously. At places like Jackson Hole, Wyoming, where the airport elevation is over 6000 feet, I've watched pilots of small jets make angry passengers wait for the temperature to cool enough to allow takeoff.

Performance Tables

Here are some extracts from the performance tables for the Cessna 414A. The choice of included data is determined by experience, which is very difficult to explain. To put it simply, a 414 cruises fast, climbs slowly, and slows down at a shocking rate after a descent. Best cruise altitudes are in the mid-teens, that is, between 14,000 feet and 20,000 feet. The aircraft can go higher, but doing so is worthwhile only in exceptional conditions.

Table 6-1 is an excerpt from the time, fuel, and distance to climb table. It shows how much fuel is required to climb from sea level to the altitude indicated, how long it will take to do so, and how far the airplane will travel. The distance traveled will be affected by wind, but the time and fuel will not.

Table 6-2 shows the cruise performance at selected power settings. Fuel consumption is directly related to power. The aircraft's engines can be set to develop, say, 55 percent power in many ways, depending on temperature, altitude, and propeller RPM, but in all cases the engine burns 154 pounds of fuel each hour.

Table 6-1. Cessna 414A Climb Performance

ALTITUDE (MSL)	TIME (MIN)	FUEL (LB)	DISTANCE (NM)
10,000	11	42	24
20,000	25	100	60

Table 6-2. Cessna 414A Cruise Performance, in Knots

ALTITUDE (MSL)	45% (130PPH)	55% (154PPH)	65% (177PPH)	75% (204PPH)
10,000	152	168	181	192
20,000	160	183	199	212

The table shows the power settings and speeds in knots at two relevant altitudes. Very few pilots cruise at power settings above 75 percent, and 65 percent is used quite commonly. These speeds are at maximum weight; the manual notes that the airplane gains 6 knots of TAS for each 1000 pound fuel reduction; this is about 1 knot per hour. Some airplanes have quite dramatic speed increase with fuel burn, although in some cases this is because they can climb to a more efficient altitude at the lighter weight.

It is also useful to look at fuel consumption per nautical mile, at least as far as making a rough estimate about a flight's feasibility. A Cessna 414A's fuel capacity is 1224 pounds. The entries in Table 6-3 are obtained by dividing the TAS by the hourly fuel consumption. The best performance is at low power and high altitude.

Planning Considerations and Route Specifics

We now have the data necessary to plan a no-wind flight *with everything working* from Goose Bay, Labrador, to Keflavik, Iceland, by way of Prins Christian Sund NDB (identifier OZN), Greenland. The total distance is 1319 NM. We will also determine the effect of wind and plan for two major contingencies: engine failure and pressurization failure.

The segment from Goose Bay to Keflavik is the longest segment of one published route from North America to Great Britain. If the aircraft can fly from Goose Bay to Keflavik, it can fly to Europe.

Most large jets use different routes, namely the North Atlantic Track System, which is adjusted daily to take advantage of the winds. But the NATS airspace is restricted to specially certified aircraft that meet stringent equipment requirements called the minimum naviga-

Table 6-3. Cessna 414A Cruise Fuel Consumption, lb/NM

ALTITUDE (MSL)	45% (130PPH)	55% (154PPH)	65% (177PPH)	75% (204PPH)
10,000	1.17	1.09	1.02	0.94
20,000	1.23	1.19	1.12	1.04

tion performance specification (MNPS) and reduced vertical separation minimums (RVSM). A Cessna 414A is unlikely to be properly equipped, and its optimum altitude is below the floor of MNPS airspace.

Planning involves simple arithmetic in overwhelming quantities. Wind calculations mean even more arithmetic. Before there were electronic computers, a navigator had to do this all by hand. This means that some compromises were made. When working by hand, it is convenient to divide the flight into zones of approximately an hour's flight time and to use the forecast winds for each zone to estimate the groundspeed. With this information, the navigator could estimate the time each zone would take, and thus the fuel consumption for the flight. Replanning at a different altitude would mean just as much work. But a computer can divide the flight into 2-mile zones[1] and calculate the fuel consumption at 10 different altitudes in less time than it takes a navigator to eat a doughnut.

In this example, we use charted points along the route as zone boundaries. The route is shown in Figure 6-1.

The published route from Goose Bay to Keflavik[2] goes from Goose Bay through LOACH (N 55° 31.0´, W 057° 05.0); the unnamed point at N 58° 00.0´, W 050° 00.0´; the Prins Christian Sund NDB (identifier OZN) located at N 60° 03.5´, W 043° 09.8´; an unnamed point at N 61° 00.0´, W 040° 00.0´; another unnamed point at N 63° 00.0´, W 030° 00.0´; a fix called EMBLA, which is 120 NM from Keflavik on the 278° radial; and finally to KEF. The coordinates of EMBLA are N 63° 28.2´, W 026° 58.9´, and KEF is at N 63° 59.2´, W 022° 36.9´.

There are no convenient alternates available enroute. Narssarssuaq is about 100 NM from Prins Christian Sund; its runway is of adequate length but there is limited fuel availability.[3] There is no instrument landing system at Narssarssuaq, so weather may make it unusable. Godthab is 200 NM farther than Narssarssuaq and has fuel, but the runway is of marginal length, and the terrain means that flight at 10,000' is risky at best. Sondre Stromfjord (Sondrestrom AB) is 380 NM from Narssarssuaq along a route with very high terrain. It is a U.S. Air Force facility, so prior permission is required (rather than "please," say "emergency"), and weather is notoriously ugly.

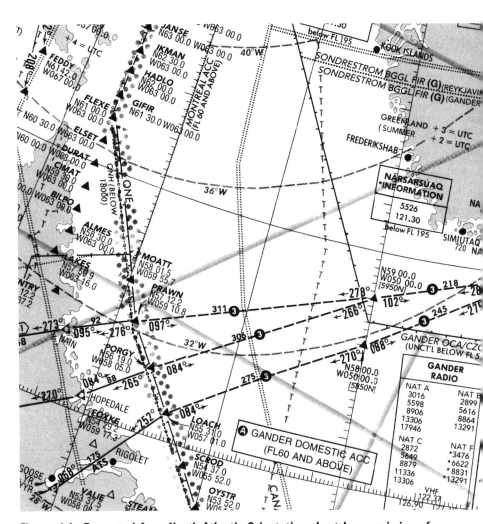

Figure 6-1. Excerpted from North Atlantic Orientation chart by permission of Jeppesen Sanderson, Inc. Copyright 1995 Jeppesen Sanderson, Inc.

Detailed Planning (No Wind)

The first line of Table 6-4 comes from the climb performance table, and the subsequent TAS figures are adjusted for weight reduction due to fuel burn. Again, this plan assumes no wind.

In doing these calculations, all times and all fuel burns are rounded up. This is the more prudent approach. In addition, the plan is to

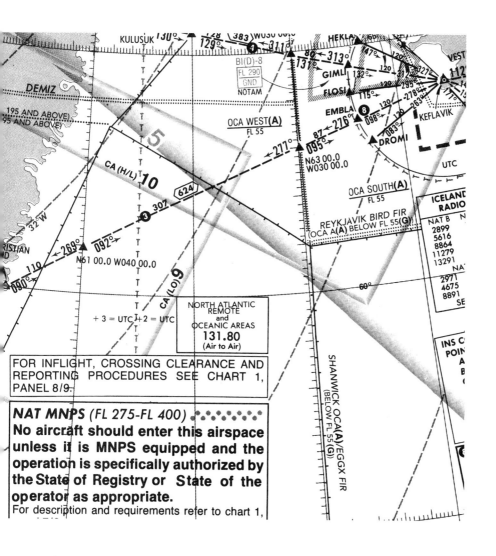

arrive over the destination at altitude; the descent and approach performance should be better, although unusual winds might make a difference. It is prudent to carry one hour of reserve fuel, although for a flight like this there are other issues to consider. In the United States, regulations require IFR flights to carry enough fuel to fly to the destination, thence to an alternate, and after that for 45 minutes. Since this flight would depart from a Canadian aerodrome, Canadian rules apply, and an alternate would be necessary (in the United States, in many cases there is no need to specify an alternate airport in a flight plan).

In this example, the total fuel burn, including a one-hour reserve, is 1231 pounds. The aircraft holds 1224 pounds, so the reserve is reduced to 123 pounds or about 57 minutes.

We can see immediately that there is a problem if there is significant headwind or crosswind, because these winds reduce groundspeed and thus increase the fuel requirements. The navigator's choices are limited. One choice is to wait for better wind. Impatient navigators may choose a different route, such as Frobisher to Sondre Stromfjord (477 NM) to Keflavik (727 NM).

Contingencies

Pressurization Failure

Even with favorable winds, we need to do a little more planning before we launch for Europe. The planned altitude of 20,000 feet is good for the airplane but not so good for people. If there is a pressurization failure,[4] the crew and passengers will be breathing air with less oxygen and will suffer *hypoxia*. How bad is this? At 20,000 feet the time of useful consciousness is about 10 minutes; anyone who stays at that altitude for longer will be too stupid or angry or euphoric to fly the airplane and will eventually lose consciousness.[5]

One cure for hypoxia is to breathe oxygen from a tank, but there may not be room in the airplane for a large-enough tank. The other cure is to descend; prolonged flight at 10,000 feet is perfectly safe for healthy individuals.

Table 6-4. Cessna 414A Flight Planning, No Wind

END FIX	GROUNDSPEED (KT)	DISTANCE (NM)	TIME (MIN)	FUEL (LB)
Top of climb	—	60	25	100
LOACH	160	115	43	94
N58 W050	161	275	102	222
OZN	163	245	91	195
N61 W040	164	110	40	87
N63 W030	165	307	112	242
EMBLA	167	87	31	68
KEF	167	120	43	93
Reserve	168	168	60	130

So, we must replan the flight at 10,000 feet. You might think that we need only plan half the flight, since if there is a problem we can continue forward if that is closer. But "half" is too vague, especially if there is wind, and it is better to plan as much as possible (the airplane can't do the whole trip at 10,000 feet). Later, we will incorporate this planning into a "howgozit" chart. For planning purposes, suppose a TAS of 152 knots at 10,000′, burning 130 lbs/hr.

Engine Failure

Another major contingency is engine failure. Almost any twin engine aircraft will fly with one engine shut down if everything is done correctly: the dead engine's propeller must be feathered, and the airplane has to be flown properly with landing gear and flaps up.[6]

The single engine service ceiling is the highest altitude at which the airplane will climb at 50 feet per minute on one engine. For a Cessna 414A, the single engine service ceiling is 20,000 feet. So if it loses an engine at 20,000 feet, it can maintain altitude, but this will require nearly full power from the operating engine. The performance figures are ugly: 135 knots TAS with a fuel burn of 195 lb/hr. This is good for 0.69 NM per pound, with no wind.

The airplane does better at 10,000 feet on a single engine. At this altitude the TAS is reduced to 122 knots, but the fuel burn is 120 lb/hr. The fuel consumption improves to 1.02 NM/lb. So, in the event of either pressurization failure or engine failure, we will have to descend to 10,000 feet.

Weather

Weather is also a consideration. Most turbulence isn't dangerous, but it hurts performance because with each bump, the airplane spends a little time pointed off course; and each correction adds a tiny amount of drag, requiring more fuel. Even worse is ice. Ice can form on an aircraft flying in visible moisture with the temperature near freezing. Ice adds drag and changes the lift characteristics of the wing, so it always hurts performance. Usually, icing occurs in a narrow range of altitudes, so it is possible to stop its accumulation by climbing or descending 2000 feet or so. But even if there is no new ice, the old ice can stay on the airplane and hurt performance. (If the aircraft can descend into warmer air, the ice will melt off, or it may sublimate in sunshine.)

The Cessna 414A is equipped with de-icing equipment and so may fly in ice, but it does not fly well. The pilot must do something to stop the ice as soon as it is noticed. An iced-up airplane over the ocean may not be able to make it to any shore.

What happens if the weather deteriorates after passing the point of no return? This has happened to U.S. Navy transports supporting Antarctic research; see (Parfit) for a thrilling story of one such flight.

The Howgozit

We now have enough data to construct a *howgozit* chart, Figure 6-2. The name is a corruption of "how goes it?" The horizontal axis of the chart shows distance, and the vertical axis is fuel consumed. Properly used, the howgozit tells the navigator how much fuel is needed for any contingency at any point in the flight.

The first step is to draw a horizontal line at the actual fuel load, in this case 1224 lbs. This line will provide a reference for fuel on board

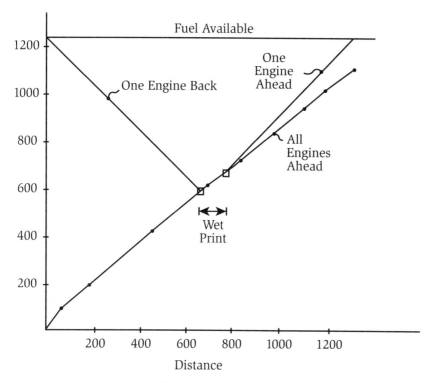

Figure 6-2. No-wind howgozit with engine failure contingency curves.

at each point: fuel on board is represented by the distance from the fuel consumed to this line.

The next step is to plot the distance and fuel burns from the table for normal cruise. This curve is slightly concave down because fuel efficiency improves as fuel is burned. In flight, a similar curve should be drawn showing how much fuel was actually used.

Now we draw two curves for each of two contingencies. The contingencies are pressurization failure and engine failure, and each contingency gets a curve for turning back and a curve for pushing on. The contingency curves are drawn backwards.[7]

First look at the "one engine ahead" curve. This is meant to tell you how much fuel you need to proceed from any point to KEF if an engine fails. In other words, you want to compare the fuel needed to the fuel available, which is represented by the distance from the normal cruise curve to the fuel line. You need nothing at KEF, so the curve starts at 1224 lbs at the right end. Since single engine flight gets 1.02 NM/lb, draw the line of slope $1/1.02 = 0.98$ through the point plotted. There is no need to extend the line past the normal cruise curve, as long as the "one engine ahead" curve is above the fuel burn curve, there is sufficient fuel on board to make it to Keflavik on one engine.

Similarly, the "one engine back" curve starts above Goose Bay at 1224 lbs and extends to the right with slope -0.98 until it meets the normal cruise curve. As long as this curve is above the fuel burn curve, there is enough fuel to make it back to Goose Bay.

This chart reveals a significant problem: the "one engine ahead" curve meets the fuel burn curve about 800 NM from Goose Bay. If you lose an engine 700 NM from Goose Bay, you will not have enough fuel to reach Keflavik. On the other hand, the "one engine back" curve meets the fuel burn curve about 650 NM from Goose Bay. So, if you lose an engine 700 NM from Goose Bay, you will not have enough fuel to reach Goose Bay, either.

This segment has a *wet print*, that is, a region in which an engine failure will leave you unable to return or to go to the destination. Any engine failure between 650 NM out and 800 NM out—an hour's exposure—means that the aircraft will run out of fuel (or be lucky enough to find Narssarssuaq).

If there is no wet print, the intersection of the normal fuel burn curve with the "one engine ahead" curve marks the *point of no return*: beyond this point you press on even if there is an engine failure.

What about pressurization failure? These curves are shallower (the aircraft gets 1.17 NM per pound) and there is no wet print. In fact, at any point between about 500 NM out to about 650 NM out, the navigator has enough fuel to either go back to Goose Bay or press on to Keflavik. In this situation weather would probably be the deciding factor.

Finally, note that the howgozit is a graphical method, so the answers it yields are approximate. There are methods for programming a computer to find the intersections that are quicker but that may not be more accurate in any significant way.

Table 6-5. Cessna 414A Flight Planning—20 Knot Tailwind

END FIX	GROUNDSPEED (KT)	DISTANCE (NM)	TIME (MIN)	FUEL (LB)
Top of climb	—	68	25	100
LOACH	180	107	36	77
N58 W050	181	275	91	198
OZN	183	245	80	174
N61 W040	184	110	36	78
N63 W030	184	307	100	217
EMBLA	185	87	28	61
KEF	186	120	39	84
Reserve	187	187	60	130

Wind Effects

It should be obvious that a Cessna 414A cannot fly between Goose Bay to Keflavik with any headwind: the no-wind plan uses all of the fuel it can carry. But a tailwind is not necessarily all good. The tailwind is great if everything is working, but the tailwind is a headwind if the navigator decides to return in the event of a pressurization or engine failure. The best way to observe this effect is to calculate another howgozit with a tailwind. See Table 6-5. We will arbitrarily use 20 knots for the whole route; more likely, the headwind would differ in each zone, and the navigator would need to calculate groundspeeds for each zone in each direction. More arithmetic.

Now the total fuel requirement, including reserves, is 1119 pounds. One could take off with this amount or with the 1224 pound maximum fuel load; the latter is prudent but the former may be financially significant.[8]

What about the contingencies? With this wind, the single engine cruise groundspeed is 172 knots at 130 lb/hr with the wind and 132 knots at 130 lb/hour against the wind. This is 1.32 NM/lb with the wind and 1.02 NM/lb against. In this example, the aircraft can always proceed to Keflavik on one engine, while it cannot turn back to Goose Bay after about 700 NM. This seems paradoxical when compared with the no-wind example: the aircraft can go farther when it fights a headwind. The explanation is the increased reserve fuel; without it, the farthest out one could go from Goose Bay and still return on one engine is about 600 NM.

NOTES

1. This is a little extreme.
2. This is an ATS route and has no formal name.
3. According to the *Canadian Forces Flight Supplement*.
4. This has happened to me at 26,000 feet.
5. Mountain climbers sometimes work at these altitudes without oxygen, although some consider them to be a lunatic fringe. Besides, they climb to these altitudes over the course of weeks, not minutes, so the body acclimates a little.
6. The proper configuration is zero sideslip.
7. My IFR students know that a flight should always be planned from the destination to the departure.
8. If nothing else, fuel is much more expensive outside the United States, so this would be 105 pounds of fuel that would not be purchased at the higher price. This practice is called *tankering fuel*.

CHAPTER 7

A Guide to Further Reading

Navigation can't be learned wholly from books, but many books explain the navigator's realm in more depth. Some of these are factual, some are fiction, some are manuals; but all have something to teach a prudent navigator who is seeking to improve his or her knowledge of the craft. And a navigator with a little experience can learn a new technique from a book and have it ready when needed in practice.

Most people have seen a math book or two in school, so there is no need to list many of them. One I find particularly appealing is (Connally et al.). There is a general movement in college mathematics called "calculus reform," which takes advantage of the availability of calculators and computers to develop the subject in a more conceptual and problem-oriented way. (Connally et al.) is not a calculus book, but it is written in the same spirit by a team who wrote my favorite calculus book.

Navigation Manuals

There are probably hundreds of books on navigation, but most of them devote only a few pages to aerial navigation. The most important is (Bowditch). Bowditch is considered the first American mathematician, and the idea of a compendium of all known information about navigational theory and practice was his. The book is regularly updated by the U.S. Navy; the most recent edition appeared in 1995. The older editions are still of interest, partly because the book concentrates on basic principles that will always be useful, and partly because older editions contain material no longer in use (for example, the lunar distance method for determining time).

The U.S. Air Force trains its navigators with (USAF). The edition I have is from 1954 and is illustrated with older airplanes such as the B-36. It discusses the use of a drift meter as an aid to dead reckoning. I have never seen a drift meter. Neither had Lindbergh on his New York to Paris flight; he estimated the winds by watching the spray on the waves below.

This is the only source I know for a comprehensive discussion on the use of radar in navigation. While most pilots study the navaids and approaches to the destination, the navigator of an attack aircraft has to depend solely on onboard sensors. Radar works as a form of pilotage: I often try to amuse other pilots by making radar pictures of local lakes and mountains.

One notable part of the manual describes polar navigation. It also contains the warning "as a navigator, ... you must have the ability to use good judgment under unusual circumstances."

The chapter "Fighter Pilot Celestial Navigation" describes methods for reducing star sights in a single pilot aircraft. Many people think this is impossible, but with proper planning and a good autopilot it can be done. This ability might come in handy on one of those night freight runs across the desert.

Another interesting treatise on polar navigation is the company training manual (SAS), which describes the instruments and techniques of polar navigation. It includes an extensive bibliography. Northern tier countries have a special interest in polar navigation, of course. This volume includes reports on SAS exploratory flights from Oslo and Stockholm to Los Angeles, Anchorage, and Tokyo by various routes. These reports note the erratic behavior of the magnetic compass and related instruments: on a flight from Thule to

Anchorage, "When we were abeam of the magnetic pole, 210 n.m. north of it, the magnetic compass had no sense of direction, the Gyrosyn compass pointed true south, the nose of the aircraft pointed to grid north and our starboard wing pointed to true north."

(Bauer) is a very complete introduction to the nautical sextant, including some history. There is very little material about the bubble sextant, which is more likely to be used in aircraft. Experience with a sextant, even standing still on dry land, will make you a better navigator: I guarantee that you will be shocked at how poorly you do at first. This will teach you to read your other instruments with more accuracy.

(Hart) is a short book designed to give wartime navigators more "depth." The author's claim was that tables had made the navigator's job very simple, but that there was still some need for the navigator to understand the basics. There is a very short chapter on aerial navigation. One wonders what Hart would say about modern GPS receivers. (Vetter) is another book from the same era. It's hard to tell if this book is aimed at the civilian pilot (there was virtually no civilian flying in the United States in 1942 due to wartime considerations) or the young Navy "nugget" (Vetter was a Navy lieutenant when this was published). Some of the weather information is out-of-date, and almost all of the electronic navigation techniques discussed are obsolete. The book is notable for almost 300 illustrations with Art Deco titles.

(Lyon) was a standard text in air navigation. It includes chapters on polar, pressure pattern, and celestial navigation, as well as all of the radio navigation techniques in use in the 1960s, instruments, and charts. There is some obsolete material about aeronautical charts (the principles are the same, but the publishing standards have changed). It is no longer in print.

(Mixter) is another of the major marine navigation textbooks that is periodically reissued.

History of Navigation

There is no better popular introduction to the history of navigation (or any other subject) than (*Encyclopaedia Britannica*). The eleventh edition was published in 1910–1911 and represented nearly all of Western civilization's knowledge at the time. A reader accustomed to modern encyclopedias will be pleasantly surprised at the depth of the its essays. Notable essays include those on Cook, parallax, and the transits of Venus.

Another classic is (Dana), which details his story of life at sea in the 1830s. Dana was a student at Harvard but was forced to take a leave of absence due to illness. He spent the time as a common seaman, and turned his ability with language to describing the seaman's life. He also describes life in California before the 1849 Gold Rush.

The book contains many navigational details. His ship had a chronometer failure early in the voyage, and relied mostly on dead reckoning and pilotage while rounding the Horn. The captain was later able to determine the time through a lunar observation, but at one point decided to heave to (that is, stop offshore) rather than approach the coast at night from an uncertain position. Aircraft have a similar but limited option: I have spent many hours circling in the clouds above mountain airports with no instrument approach, waiting for better conditions. The limit here is that you have to stop circling while you still have enough fuel to get home.

(Byrd) wrote several books about exploring Antarctica; all are carefully written and include maps, photos, and extracts from flight logs. See Chapters XIII and XIV for details on the polar flights.

Polar exploration has always been controversial, and the navigational aspects are especially so. Charges and countercharges about whether so-and-so really reached his avowed destination continue to this day (see (Portney), pp. 11–30.) Amundsen, however, was a methodical explorer, and skied a 1 kilometer box around the South Pole in order to be sure that he had reached it. Amundsen's navigation is described in (Huntsford).

(Wohl) is a historian and a pilot, and he has taken a serious look at the effects that early aviation had on poetry and art. The book contains many high-quality reproductions. There are amusing anecdotes about authors such as Franz Kafka. He also discusses the use of aviation as a political metaphor, especially with regard to the Soviet

Union and Fascist Italy. The book has an academic tone but is nonetheless captivating.

(Wright) is a comprehensive history of aerial navigation (including airships and balloons), written by a historian who had been a navigator in the U.S. Air Force and later served as the director of NASA's history office. This book is unfortunately out of print.

(Li and Shiran) discuss some of the navigational techniques known in classical China.

(Williams) is a lavishly illustrated and comprehensive history of navigation.

(Post and Gatty) discuss Wiley Post's pioneering flights in the *Winnie Mae.* Gatty was a pioneer aerial navigator who invented and improved several instruments. This edition includes reproductions of Gatty's logs and discussions of some of the difficulties encountered in overflying unfamiliar and unmapped terrain. The book is also remarkable for some of the egregious errors (from a modern flight instructor's perspective) Post made. For example, he clearly misunderstood the difference between indicated and true airspeed, and he used his rudder trim to compensate for drift!

There is a strong argument that the leader in long-range aerial navigation was Pan American during the era of the clippers. These were the first routine ocean crossings, and Pan Am had its own navigational methods and schools for training navigators. Some of this is discussed in (B. Taylor).

The great dirigibles, such as the Graf Zeppelins, the *Norge* and the *Hindenberg*, were also pioneers of long-range flight over remote terrain. (Toland) describes them and their navigational problems. Especially gripping is the flight of the *Norge*, which carried Amundsen, Ellsworth, and crew across the North Pole from Norway to Alaska. The navigator, Riiser-Larsen, was forced to take observations from atop the fast-moving ship in frigid Arctic air.

(Portney) has more than 20 essays, each presenting an interesting historical navigational problem with its solution. These aren't the artificial problems one sees in the magazines, where the reader is supposed to use this year's almanac to reduce a sight taken in 1732; these problems are the real thing. Personalities, in addition to Byrd and Peary, include Columbus, Lewis and Clark, and Earhart.

Judgment and the Navigator

Aerial navigation today is more and more automated, and according to (Billings) and others our understanding of automation and its limits is incomplete. The navigator (or pilot who navigates as one part of managing a highly complex and automated aircraft) needs more awareness of these issues and how they affect flight. (Perrow) discusses how system design can lead to accidents. There is a chapter on aircraft incidents, but the real lesson is that since a navigator must interact with complex systems, the errors that complex systems induce should be studied.

(Callahan) spent 76 days adrift at sea in a lifeboat. His book describes some of the primitive improvised navigational methods he used. (Rickenbacker) wrote about spending 21 days at sea after ditching a B-17 due to navigational errors.

(Gann) is a thrilling book written by a retired airline pilot. Notable sections include early flights to South America and Greenland. He gets to the heart of the navigator's role and the illusions that can kill. This is well worth re-reading. (Someday I will try to assign this as reading to the new charter pilots I train.)

(Hutchins) is a cognitive psychologist (and pilot) who studied the teaching and practice of navigation aboard U.S. Navy vessels and developed a theory of group intelligence based on the findings. This book discusses many of the psychological aspects of navigation: the design of procedures, the use of tools, communication, and the like. It is an academic book, but there is a lot of interesting reading, including a new analysis of Micronesian navigational techniques.

(Raban) sailed his yacht from Seattle to Juneau via the Inland Passage. This book tells the story of that journey, skillfully weaving Native and European history, a biographical sketch of Captain Vancouver, navigational lore, and personal searching. He used a mixture of GPS navigation and more traditional techniques, including celestial navigation. One would think that pilotage would be the most important technique during a passage completely within sight of land, but the currents in the Inland Passage are very tricky and he still found GPS useful.

Navigation requires physical endurance: read what (Worsley) says about navigating Shackleton to South Georgia Island in an open boat, or what (Toland) wrote about the *Norge*, or (Byrd).

Regulations and the Like

The Airman's Information Manual is an FAA document, available from commercial publishers, which describes the air traffic control system including aids to navigation, radio techniques, lighting and marking, weather services, and the like. It is updated on a 56-day cycle. Every pilot should own a recent version. The ICAO equivalent is the AIP issued by most countries.

Equipment and Technique

The best introduction to the engineering aspects of navigation is (Kayton and Fried). It is an engineering textbook, written for engineers and scientists. It has complete details on the inner workings of all major navigation systems, including GPS, INS, VOR, DME, NDB, celestial, and radar. Here you can learn which frequencies GPS uses, how many aircraft a DME station can service, the weight of a typical INS unit, and the like.

(Landes) is a comprehensive history of timekeeping, written by an economic historian and collector of clocks and watches. It has a comprehensive and fair-minded history of the development of the marine chronometer. It can be contrasted with (Sobel), a very popular history of the chronometer and the personalities involved in its acceptance as a navigational instrument. Sobel tries to make Harrison the hero and Maskelyne the villain, but modern techniques use both Harrison's chronometer and Maskelyne's Ephemerides. (Sobel) was made into a TV movie starring Jeremy Irons.

The least-expensive almanac is (Paradise Cay), based on data supplied by the governments of the United States and Great Britain. It catalogs complete tables for Aries, Mars, Jupiter, Saturn, the Sun, and

Moon, including everything one needs to interpret celestial sights. It also includes a tabular method of sight reduction, extensive documentation, and a section providing details on calculated solutions.

The most comprehensive almanac is *The Air Almanac* (U.S. Naval Observatory). It includes detailed star identification diagrams showing the field of view of a bubble sextant, and also has rough diagrams of the stars and planets designed to ease the choice of which bodies to observe.

(Sacchi) flew dozens of light airplanes across the oceans. Some of the information here is out-of-date (there is no mention of Global Positioning System, for example), but many of the problems she faced remain. She places a little bit too much faith in her "guardian angel" for my taste.

The most comprehensive introduction to the equipment used in airplanes is (Illman).

Maps, Charts, and Geodesy

Everyone is familiar with charts and maps, but most are less familiar with the mechanics of constructing charts. This is a subtle area because the shape of the Earth is so subtle.

For maps and mapmaking, (Snyder) considers everything from Rennaissance to modern computer maps. This book includes some mathematical derivations and a large number of striking examples. (Wilford) is an interesting history of cartography, with up-to-date information on satellite mapping; it is pleasant reading while being technically correct. (Snyder) is the more mathematical of the two books.

(Heiskanen and Moritz) is a technical introduction to the subject of geodesy. It requires knowledge of advanced calculus.

Navigation and Literature

(Wohl), mentioned above, outlines the effect that the advent of aviation had on literature, poetry, and the plastic arts. Fewer books mention navigation extensively, although I imagine that the fear of being lost is an essential component of the human condition. The books of Anne Morrow Lindbergh come to mind, especially *North to the Orient*, as do the books of her close friend Antoine de Saint-Éxupéry, who wrote eloquently about flight. In the short novel (Saint-Éxupéry), an airmail pilot is lost. Its characters are notable (and not always likable); the best navigator was the protagonist's wife, who was able to keep a dead reckoning of his fatal flight in her head.

Magazines and Journals

Navigators and engineers still develop new techniques. New engineering and commercial ventures are announced in *Aviation Week and Space Technology*, while pilot techniques are often addressed in its sister publication, *Business and Commercial Aviation*. New gadgets for small planes are described in *Flying* and AOPA *Pilot*. The publications of the Institute of Navigation contain a wide variety of articles on navigation, ranging from personal histories of developers to arcane models of subcentimeter GPS positions. *Ocean Navigator* includes many interesting articles on navigation at sea.

The Navigational Stars

This appendix contains the declination and sidereal hour angle of each of the 57 navigational stars, from the *Air Almanac* of 1999. A prudent navigator would use a current almanac.

STAR	SHA (° ′)	N/S	DECLINATION (° ′)
Alpheratz	357 55	N	29 05
Ankaa	353 26	S	42 19
Schedar	349 53	N	56 32
Diphda	349 07	S	17 59
Achernar	335 35	S	57 14
Hamal	328 13	N	23 27
Acamar	315 27	S	40 19
Menkar	314 26	N	4 05
Mirfak	308 56	N	49 51
Aldebaran	291 02	N	16 30
Rigel	281 22	S	8 12
Capella	280 50	N	46 00
Bellatrix	278 44	N	6 21
Elnath	278 26	N	28 36

STAR	SHA (° ′)	N/S	DECLINATION (° ′)
Alnilam	275 57	S	1 12
Betelgeuse	271 13	N	7 24
Canopus	264 01	S	52 42
Sirius	258 43	S	16 43
Adhara	255 21	S	28 58
Procyon	245 11	N	5 14
Pollux	243 41	N	28 02
Avoir	234 22	S	59 31
Suhail	223 00	S	43 26
Miaplacidus	221 42	S	69 43
Alphard	218 07	S	8 39
Regulus	207 55	N	11 58
Dubhe	194 05	N	61 45
Denebola	182 45	N	14 35
Gienah	176 03	S	17 32
Acrux	173 21	S	63 06
Gacrux	172 13	S	57 07
Alioth	166 30	N	55 58
Spica	158 43	S	11 09
Alkaid	153 07	N	49 19
Hadar	149 03	S	60 22
Menkent	148 20	S	36 22
Arcturus	146 06	N	19 11
Rigil Kentaurus	140 07	S	60 50
Zubenelgenubi	137 17	S	16 02
Kochab	137 20	N	74 10
Alphecca	126 20	N	26 43
Anteres	112 40	S	26 26
Atria	107 51	S	69 01

STAR	SHA (° ′)	N/S	DECLINATION (° ′)
Sabik	102 25	S	15 43
Shaula	96 37	S	37 06
Rasalhague	96 17	N	12 34
Eltanin	90 51	N	51 29
Kaus Australis	83 58	S	34 23
Vega	80 46	N	38 47
Nunki	76 12	S	26 18
Altair	62 19	N	8 52
Peacock	53 36	S	56 44
Deneb	49 39	N	45 17
Enif	33 58	N	9 52
Al Na'ir	27 57	S	46 58
Fomalhaut	15 36	S	29 38
Markab	13 49	N	15 12

Aviation Week and Space Technology, July 28, 1997.

Bate, Roger R., Donald D. Mueller, and Jerry E. White, *Fundamentals of Astrodynamics*. NY: Dover (1977).

Bauer, Bruce, *The Sextant Handbook*, Camden, ME: International Marine (an imprint of McGraw-Hill) (1995).

Billings, Charles E., *Aviation Automation : The Search for a Human-Centered Approach*. Mahwah, NJ: Lawrence Erlbaum Associates (1996).

Bowditch, Nathaniel, *The American Practical Navigator*, U. S. Navy Hydrographic Office Publication HO. 9, 1962. (Most recent edition 1995.)

The British Admiralty, *Admiralty Navigation Manual*. London: H. M. Stationery Office (1938).

Byrd, Richard E., *Little America*. NY: G. P. Putnam's Sons (1930).

Callahan, Steven, *Adrift*. Boston: Houghton-Mifflin (1986).

Chernicoff, S., *Geology*. NY: Worth (1995).

Connally, Eric, et al., *Functions Modelling Change: A Preparation for Calculus*. NY: Wiley (2000).

Cotter, Charles H., *A History of Nautical Astronomy*. NY: American Elsevier (1968).

Dana, Richard Henry, Jr., *Two Years Before the Mast*. NY: Harper Perennial Classics (1954).

Encyclopaedia Britannica, Eleventh Edition.

Farmer, Gene, *First on the Moon, A Voyage with Neil Armstrong, Michael Collins, and Edwin E. Aldrin, Jr.* Boston: Little, Brown (1970).

Featherstone, Will, and Richard B. Langley, "Coordinates and Datums and Maps! Oh My!" *GPS World*, January 1997.

Federal Aviation Administration, *The Airman's Information Manual*. Published every 56 days.

Flying, December 1983.

Gann, Ernest K., *Fate is the Hunter*. NY: Simon and Schuster (1961).

Hart, M. R., *How to Navigate Today*. NY: Cornell Maritime Press (1944).

Heiskanen, Weikko A., and Helmut Moritz, *Physical Geodesy*. Graz, Austria: Institute of Physical Geodesy, Technical University Graz (1999).

Hunerfelt, Sigurd, *The Earth According to WGS-84*. Available at http://home.online.no/~sigurdhu.

Huntford, Roland, *Scott and Amundsen*. NY: Atheneum (1984).

Hutchins, Edwin, *Cognition in the Wild*. Cambridge, MA: MIT Press (1995).

Illman, Paul E., *The Pilot's Handbook of Aeronautical Knowledge*. NY: McGraw-Hill (1999).

Kaplan, George H., "Determining the Position and Motion of a Vessel from Celestial Observations," *Navigation*, vol. 42, no. 4, Winter 1995–1996.

Kayton, Myron, and Walter A. Fried, *Avionics Navigation Systems*, Second Edition. NY: John Wiley and Sons (1997).

Landes, David S., *Revolution in Time*. Cambridge, MA: Harvard (1983).

Li You and Du Shiran, *Chinese Mathematics: A Concise History*. Oxford University Press (1987).

Lindbergh, Anne Morrow, *North to the Orient*. NY: Harcourt (1996).

Lindbergh, Charles, *The Spirit of Saint Louis*. NY: Scribner (1998)

Lyon, Thoburn, *Practical Air Navigation*, Commercial Edition of Civil Aeronautics Bulletin No. 24. Eighth Edition. Annapolis: Weems (1960).

Mixter, G. W., *Primer of Navigation* New York: D. Van Nostrand (1943).

National Geodetic Survey, *Data Sheets*. Available through http://www.ngs.noaa.gov.

National Transportation Safety Board, Accident Report SEA95LA039. Available at http://www.ntsb.gov/aviation

Paradise Cay, *The Nautical Almanac*. Commercial Edition. Arcata, CA: Paradise Cay Publications. (Annual.)

Parfit, Michael, *South Light*. NY: Macmillan (1985).

Perrow, Charles, *Normal Accidents: Living with High Risk Technologies*. Princeton, NJ: Princeton University Press (1999).

Portney, Joe, *Portney's Ponderables*. Woodland Hills, CA: Litton Systems — Guidance & Control Systems Division (2000).

Post, Wiley, and Harold Gatty, *Around the World in Eight Days*. NY: Orion Books (1989). (Originally published by Rand McNally c 1931.)

Raban, Jonathan, *Passage to Juneau: A Sea and Its Meanings*. NY: Pantheon Books (1999).

Rickenbacker, Captain Edward V., *Seven Came Through*. Garden City, NY: Doubleday, Doran and Company (1943)

Sacchi, Louise, *Ocean Flying*, NY: McGraw-Hill (1979).

Saint-Éxupéry, Antoine de, *Vol de Nuit* (English version *Night Flight*). Harcourt Brace (1986).

Scandinavian Air System, *Polar Navigation*. Scandinavian Air System Company Training Library Operations Series No. 100.56.

Science Magazine, October 31, 1884.

Sherry, Frank, *Pacific Passions: The European Struggle for Power in the Great Ocean in the Age of Exploration*. NY: William Morrow and Company (1994).

Slocum, Joshua, *Sailing Alone Around the World*. NY: Dover (1956).

Snyder, John P., *Flattening the Earth: Two Thousand Years of Map Projections*. Chicago: University of Chicago Press (1993).

Sobel, Dava, *Longitude*. NY: Penguin (1995).

Taylor, Barry, *Pan American's Ocean Clippers*. Blue Ridge Summit, PA: TAB Aero (1991).

Taylor, E. G. R., *The Haven Finding Art*; *a History of Navigation from Odysseus to Captain Cook*. NY: American Elsevier (1971).

Toland, John, *The Great Dirigibles: Their Triumphs and Disasters*. NY: Dover (1992).

Transport Canada, *Canadian Forces Flight Supplement*. Issued every 56 days.

USAF, *Air Navigation*. Air Force Manual 51-40. March 1955.

U.S. Naval Observatory, *The Air Almanac*. Issued by The Nautical Almanac Office, United States Naval Observatory, (Annual).

U.S. Navy Hydrographic Office, *HO 214, Tables of Computed Altitude and Azimuth*. Washington: U.S. Navy Hydrographic Office (1936). Reissued as HO 229.

Vetter, Ernest G., *Visibility Unlimited: An Introduction to the Science of Weather and the Art of Practical Flying*, Second Edition. NY: William Morrow and Company (1942).

Wilford, John Noble, *The Mapmakers*. NY: Knopf (1981).

Wilkins, Sir George H., *Flying the Arctic*. NY: Grosset and Dunlop (1928).

Williams, J. E. D., *From Sails to Satellites*. Oxford: Oxford University Press (1992).

Wohl, Robert, *A Passion for Wings: Aviation and the Western Imagination, 1908–1918*. New Haven, CT: Yale University Press (1994).

Wolper, James S., "(Celestial) Navigation on a Riemannian Earth," *SIAM Review* **30** (1988), 498–501.

Worsley, F. A., *Shackleton's Boat Journey*. NY: W. W. Norton and Company (1987).

Wright, Monte Duane, *Most Probable Position: A History of Aerial Navigation to 1941*. Lawrence, KS: The University Press of Kansas (1972).

M

Mach number, 184
map shift, 205
matrix, 103
Mercator chart, 55, 124, 141
meridian angle, 128
meridional parts, 141
meter, 16
minimum navigation performance specification, 226–227
minutes, 16
missed approach point, 212
Morristown Museum, 120
moving map, 203, 206

N

nautical mile, 16
navaids, 114, 177
NDB, 191–194
normal, 20, 96–97, 99–101
North Atlantic Track System,
   226
NOTAM, 158

O

oblate, 61, 68
omni bearing selector (OBS),
   197
opposite, 22
orientation, 6
overlay, 197

P

parameter, 95
parametrized, 67
pilot's operating handbook
   (POH), 224
pilotage, 170, 177
pitot tube, 179–180
plane, 4, 104–106
point of no return, 217–218, 234
point-slope, 95
points, 14
polar coordinates, 38–39

prime meridian, 43
project, 36
projective geometry, 37
pseudolite, 158
pseudorange, 152–153
Pythagorean formula, 7

Q

QFE, 121
QNE, 121
QNH, 121
QTE, 121

R

radial line, 20
radian measure, 88
radians, 14
radio magnetic indicator (RMI),
   192
radio range, 195
radius, 2
rated, 120
receiver autonomous integrity
   monitoring (RAIM), 156–157
reduced vertical separation
   minimums, 227
repeaters, 47
residual, 206
resultant, 77
rhumb line, 55, 136
right-hand rule, 6, 103

S

scalar, 78–79
scalloping, 199
seal level, 71
seconds, 16
selective availability, 157
sidereal hour angle, 116, 249
single engine service ceiling,
   231
slant range error, 202
slope-intercept, 94
special use, 209
speed line, 142–143

About the Author

James S. Wolper is Professor of Mathematics at Idaho State University and a professional pilot and flight instructor. He is also on the computer science and environmental engineering faculties at ISU. He has a bachelor's degree in mathematics from Harvard University and a Ph.D. in mathematics from Brown University. His research has been supported by NASA and by the U.S. Air Force Office of Scientific Research.

An airline transport-rated pilot and flight instructor with single-engine, multi-engine, and instrument ratings, he has more than 2900 hours of flight experience, including aerial fire suppression, air ambulance flights, and more than 800 hours as a flight instructor for primary and advanced students.

He is the author of numerous technical articles and reviews. This is his first book.